HUMMINGBIRDS

HUMMINGBIRDS

A Life-size Guide to Every Species

MICHAEL FOGDEN, MARIANNE TAYLOR, and SHERI L. WILLIAMSON

Foreword by Pete Dunne

HARPER
DESIGN

An Imprint of HarperCollins Publishers

Hummingbirds: A Life-size Guide to Every Species

Copyright © 2014 by Ivy Press Ltd.

HarperCollins books may be purchased for educational, business, or sales promotional use. For information, please email the Special Markets Department at SPsales@harpercollins.com.

Published in the United States and Canada in 2014 by:
Harper Design
An Imprint of HarperCollins*Publishers*
10 East 53rd Street
New York, NY 10022
Tel: (212) 207-7000
Fax: (212) 207-7654
harperdesign@harpercollins.com
www.harpercollins.com

Distributed throughout the United States and Canada by:
HarperCollins*Publishers*
10 East 53rd Street
New York, NY 10022
Fax: (212) 207-7654

This book was produced by:
Ivy Press
210 High Street, Lewes
East Sussex BN7 2NS, UK

CREATIVE DIRECTOR Peter Bridgewater
PUBLISHER Susan Kelly
EDITORIAL DIRECTOR Tom Kitch
ART DIRECTOR James Lawrence
COMMISSIONING EDITOR Kate Shanahan
EDITORS David Price-Goodfellow & Susi Bailey
CONTRIBUTING EDITOR Rob Yarham
DESIGNER J.C. Lanaway
MAP ILLUSTRATIONS BirdLife International

Library of Congress Control Number: 2013952639

Color origination by Ivy Press Reprographics

ISBN: 978-0-06-228064-0

Printed in China

First printing, 2014

Contents

Foreword

At no other time in human history have more people lived in closer proximity to and in greater intimacy with birds. Birds today are an integral part of our suburban landscape. They have even infiltrated urban centers, making themselves at home in habitats designed solely for our species. This avian integration was made clear to me one morning in 2013, when we, the members and staff of the New Jersey Audubon Society, spent the morning bird-watching in Central Park in New York City, spotting dozens of brightly colored warblers in a few hours. It was prime time for northbound songbird migration, and among birders a bird-walk through the meadows called the "Ramble" is a rite of spring. Later, at the apartment of two of our supporters, I inquired about the couple's "yard list" (i.e., the total number of bird species tallied in or from their yard). After they explained that as residents of the fourth floor they had no yard, I observed that they did have a fire escape and then asserted somewhat brashly that I would find 30 species of birds as a down payment on their Manhattan yard list.

Supporting my conviction were the images I saw through my Zeiss spotting scope as I scanned the East River. I found the promised number, and the event heightened my understanding of how what has lagged is human awareness, not appreciation, of the world of birds.

If asked, most people will tell you they "like" birds. Birds have integrated themselves into our religions, fables, and literature. The US Fish and Wildlife Service estimates that more than 17.3 million people actively watch birds in North America. The adaptive specializations that allow hummingbirds to hover, fly backward—even inverted (it is reported)—make them a favorite bird group among bird-watchers. Their jewel-like plumage, coupled with their partiality to nectar-bearing flowers, endears them to gardeners who have come to think of hummingbirds as animate blossoms, offering a dynamic element to landscape design.

Our appreciation of birds might be broad, but it is not uniform. Some bird groups simply excite us more than others. Penguins and hawks are hot-button bird groups; so too are hummingbirds. If few hummingbirds have been the focus of documentaries or found their way onto football jerseys—as have assorted birds of prey—I submit that few penguins are featured on fine china, yet hummingbirds are a perennial porcelain favorite.

The late writer/ornithologist Sally Spofford once told me about a pair of Blue-throated Hummingbirds that overwintered at her home near the Chiricahua Mountains. Although she put out an ample number of hummingbird feeders, rival males pummeled each other in an effort to dominate the food source. During one spirited contest, Sally observed a bird jumping up and down on his rival, driving him into the snow. But because hummingbirds have very short legs, neither bird was able to bring its lance-like bill to bear.

My wife, Linda, and I once camped for a week in Cave Creek, Arizona. The morning of our departure our hummingbird feeder was stowed on the floor of our RV and a

Blue-throated Hummingbird positioned himself between us and the exit, refusing to surrender the right-of-way until Linda offered up one last sip. While most birds flee from us, hummingbirds approach us, even sipping sugar water out of our hands if it is offered. They seem to trust us, and we repay them with affection.

My most memorable encounter with a hummingbird involves one of the authors of this book. While leading a workshop for teenage birders, Linda and I enlisted Sheri Williamson to treat our group to a banding demonstration, which required special permits. Before releasing a Magnificent Hummingbird, Sheri asked if any member of our group wanted to hold the bird. A young woman from New York City accepted the challenge and before you could say *Eugenes fulgens*, the onyx-breasted bird was lying tranquilly in the girl's hand. Never one to miss a teachable moment, Sheri told the keeper that "probably fewer than a hundred people on earth have ever held an adult male Magnificent Hummingbird before." When the girl's hand began to tremble, the bird righted itself and flew off. I don't know whether that encounter changed the young camper's life, but I am confident she never forgot it.

Given the captivating nature of hummingbirds, I am brought to wonder why it has taken so long for this book to have been published. It is the most comprehensive monograph on

hummingbirds since John Gould's epic work Hummingbirds was published in 1881. It is also unique in that it has the birds featured at their actual size. Like Gould's book, it is destined to be a classic—a book you will turn to for visual delight and also as an authoritative resource every time you encounter one of these captivating birds.

If you find yourself looking for a challenge, consider seeking out all the hummingbird species found in the New World. Given the eco-tourist infrastructure that has evolved across the Americas, it is possible to view in the field most —if not all—the species in this book. But if your travels take you no more than a few paces from the table that showcases your copy of Hummingbirds, then, thanks to the skill of photographers Michael and Patricia Fogden and the text crafted by ornithologist Sheri Williamson and writer/naturalist Marianne Taylor, you already have the world of hummingbirds at your fingertips. Consider this a down payment on the 10,000 bird species apportioned across the planet, awaiting your discovery and appreciation.

So far no one has seen all of the planet's birds. You could be the first—and your adventure begins at the nearest door, window, or fire escape. No matter where you are at this very moment, you are within sight or earshot of a bird. Now there are only 9,999 species and a lifetime of encounters to go.

Pete Dunne

AUDUBON CAPE MAY BIRD OBSERVATORY, NEW JERSEY   *November 5, 2013*

Within my Garden, rides a Bird
Upon a single Wheel—
Whose spokes a dizzy Music make
As 'twere a travelling Mill—

He never stops, but slackens
Above the Ripest Rose—
Partakes without alighting
And praises as he goes

FROM "WITHIN MY GARDEN, RIDES A BIRD " (ca.1862)—EMILY DICKINSON

Introduction

Hummingbirds are tiny, captivating, vibrant, and pugnacious. They include the smallest birds in the world and many that are sublimely beautiful, with glittering, iridescent plumage ornamented with showy tufts and plumes. They have unique flying skills and a rare ability to enter a state of torpor. The smallest species have the highest metabolic rate of any warm-blooded animal and spend much of their life on a knife edge, often within hours of death by starvation. If they were any smaller, they would be unable to produce energy quickly enough to survive. Hummingbirds have extraordinary appeal, and their extravagant names, full of allusions to jewels, fairies, and sunshine, give an idea of the way in which they have captured the imagination. Fiery Topaz, Crowned Woodnymph, Empress Brilliant, Shining Sunbeam, and Long-tailed Sylph are good examples, as is the ancient Aztec name *huitzitzil* (rays of the sun). The Mexican *chuparosa* (rose-sucker) and Brazilian *beija flôr* (flower-kisser) are picturesque descriptions of the birds' feeding behavior, while the Cuban *zum-zum* is an apt description of how they sound in flight.

The beauty of hummingbirds appealed to artists, authors, and poets in the Western world, as expressed in Emily Dickinson's poem "Within my Garden, rides a Bird ."

This fascination proved deadly for the birds, however, as it became fashionable in the mid-nineteenth century to use their feathered skins to decorate women's hats. In South America, millions of hummingbirds were slaughtered and then exported to Europe. Many specimens found their way into museums, including some species described in this book that have never been seen since.

Today, hummingbirds continue to captivate us, but luckily for the species they now attract tourists rather than trappers. The growth in hummingbird tourism has escalated thanks to the rapid proliferation of hummingbird feeders in locations all over North, Central, and South America. It has never been easier to see rare or spectacular hummingbirds at close quarters than it is today—go to the right feeders in the right country, and it is even possible to see such iconic species as the Crimson Topaz, Gorgeted Sunangel, Wire-crested Thorntail, Marvelous Spatuletail, Sword-billed Hummingbird, Booted Racket-tail, and Velvet-purple Coronet.

According to the most recent taxonomic classifications, hummingbirds currently include 338 species in 105 genera. The majority are distributed through the New World tropics, with diminishing numbers to both the north and south. The Rufous Hummingbird breeds as far north as Alaska and the Green-backed Firecrown as far south as Tierra del Fuego. Few families of birds have representatives in such extreme environments, ranging from steamy equatorial rainforests to hot, dry deserts and freezing mountain peaks.

OPPOSITE The male Violet Sabrewing, here visiting a spiral ginger, is a large, spectacular hummingbird. It mainly traplines but is aggressive and dominant if it needs to be.

Evolution, systematics & taxonomy

Based on morphological characters, hummingbirds (family Trochilidae) have traditionally been thought to share a common ancestor with swifts (family Apodidae). Shared characteristics include the wing structure, which differs from that of all other birds. The arm bones of swifts and hummingbirds are reduced in length, and the hand bones are lengthened—the wing is mostly "hand." They also have a unique ball-and-socket joint where the shoulder girdle is connected to the very large breast bone. Swifts and hummingbirds share exceptional powers of flight, and experts have often debated whether these similarities are due to convergence or to common ancestry. Today, the reality of the relationship is generally accepted, supported by genetic studies and other biochemical evidence. It is now thought that swifts and hummingbirds diverged from their common ancestor as long ago as the beginning of the Paleocene epoch, 65 million years ago, though the main radiation is thought to have taken place in the Miocene, 12 to 13 million years ago.

It was thought that hummingbirds originated in South America and have always been confined to the New World. However, in 2004, two 30-million-year-old German fossils were

identified as primitive hummingbirds. They were named *Eurotrochilus inexpectatus*, which translates as "unexpected European hummingbird."

The long-standing division of hummingbird species into two apparently natural groups, the hermits (Phaethornithinae) and the typical hummingbirds (Trochilinae), is now thought to be incorrect. Recent genetic studies indicate that four species—the topazes and jacobins—represent the oldest split from all other hummingbirds and are now classified as the subfamily Florisuginae. The hermits continue as a natural subfamily, with 36 species allocated among six genera.

The remaining 298 typical hummingbirds, which formerly comprised the bulk of the original Trochilinae, are now split among four subfamilies: the Polytminae, Lesbiinae, Patagoninae, and Trochilinae. Currently, these subfamilies include 97 genera, 49 of them monotypic, a good indication that relationships are still not fully understood. Ongoing revision of some genera, notably *Amazilia* and *Hylocharis*, is likely to involve major changes.

BELOW Male White-necked Jacobins are among the most elegant of hummingbirds. They spend much time pursuing tiny insects, twisting and turning in the air in an intricate dance.

Color & iridescence

Unlike such attractive birds as American warblers, tanagers, and orioles, whose brilliant colors appear the same whatever the position of the viewer, the iridescent colors of hummingbirds change with the angle at which they are observed. Colors are often fleeting, and even appear black in poor light. Non-iridescent colors are produced by chemical pigments in the barbules of the feathers, which absorb some wavelengths of white light and reflect the rest, which the eye sees as the complementary color. On the other hand, the iridescent colors of hummingbirds are structural rather than chemical in nature, and are caused by interference.

Interference can perhaps be understood by considering the rainbow colors seen on a thin film of gasoline on a puddle of water. Light is reflected by both the water and gasoline surfaces but, because of the distance it travels through the thin film of gasoline, light of a particular wavelength reflected by the water is out of phase with light of the same wavelength reflected by the gasoline. The out-of-phase wavelengths interfere with one another and cancel each other out, the result being that the remaining wavelengths no longer combine to make white light and are seen as a color. The surface of the gasoline film appears rainbow-colored because the distance light travels

through the film, and hence the color that is produced, varies with the viewing angle. Since the wavelengths of light are very short, interference occurs only with very thin films. In hummingbirds, suitable films occur in the form of stacks of microscopically thin platelets in the outer layer of the barbules.

The dazzling iridescence seen on the crown, gorget, and breast of many hummingbirds is highly directional, appearing blackish when viewed from the side. It is caused by the mirror-like surfaces of the color-producing barbules. In the case of the vivid colors of the gorget, the "mirrors" concentrate the color so that it can be seen only from directly in front, as it would

be seen by a territorial rival in a head-on confrontation. The brilliant gorgets of Gorgeted Sunangel are good examples (see opposite). Less brilliant colors result if the "mirrors" are curved so that the iridescence is scattered in all directions. For example, the iridescent greens seen on the body plumage of many hummingbirds, including the Sparkling Violetear are less intense and can be seen from almost any angle (see above).

Iridescent colors are rarely found in flight feathers, because the barbules that produce iridescence are modified and twisted in a way that weakens the feather structure. Iridescent flight feathers therefore do not stand up well to the stresses involved in the aerial acrobatics of a hummingbird. Nevertheless, iridescent flight feathers are found in a few species, notable examples being the Purple-throated Carib (page 80) and the Great Sapphirewing (page 174).

OPPOSITE & ABOVE The iridescent throat of the male Gorgeted Sunangel (opposite) is visible only from almost directly in front while the iridescence of the Sparkling Violetear (above) is apparent from all directions. Note the yellow pollen on the Violetear's forehead.

Flight

The ability to hover for prolonged periods while feeding at flowers sets hummingbirds apart from all other groups of birds. Hummingbirds alone can generate lift from both the forward and backward strokes of their wings, made with the wings fully extended. In other birds, the backstroke is a recovery stroke, made with partially folded wings, so that it generates no lift.

The hummingbird wing differs from that of a "normal" bird in being rigid except at the shoulder joint, which is flexible, allowing movement in all directions.

The wing can rotate about its axis and change angle by nearly 180 degrees. It is this rotation that allows the hummingbird to invert its wings with each backstroke, enabling the front edge of the wing to lead at all times and so generate lift from both strokes. Smaller changes in wing angle allow the hummingbird to make minute adjustments in its hovering position in all directions. Because they are articulated only at the shoulder, the wings of hummingbirds function more like those of insects than those of other birds. It is noteworthy, therefore, that the precision flying of hummingbirds is

matched or surpassed only by such insects as dragonflies, hoverflies, and hawkmoths.

As a hummingbird hovers in front of a flower, its rapidly beating wings push it backwards and forwards with each beat. In a large hummingbird with slow wing beats, the jerky movement is clearly visible. In small hummingbirds with wings beating 50 to 80 times per second, the jerkiness is smoothed out. Rapidly beating wings increase the stability of hovering.

Within each species, including hummingbirds, birds flap their wings at a more or less constant rate. This rate follows the same laws that govern

the movement of a swinging pendulum, which has a natural frequency depending on its length. A long pendulum swings more slowly than a short one. Similarly, long wings have a natural flapping frequency that is slower than that of short wings. Birds can and do change the rate at which they flap their wings, perhaps to elude a predator or during a courtship flight, but only for short bursts.

It is commonly supposed that hummingbirds flap their wings faster than other birds. This is generally true, but only because they are smaller and have shorter wings than most other birds. Surprisingly, when the larger hummingbirds are compared with birds of equal size, they are found to have slower wing beats. The Giant Hummingbird, for example, flaps its wings only 10 times per second, slower than the 14 times per second of the much bigger mockingbirds.

Feeding

All the most notable attributes of hummingbirds —their diminutive size; long, slender bill; acrobatic flying ability; iridescent colors; and social behavior—stem from their dependence on nectar, which they rely on as an energy source and consume while hovering at flowers. The flower connection is the result of a long history of coevolution between hummingbirds and plants.

Although they depend on flowers for their energy, all hummingbirds also need protein, fat, and other nutrients, which they get from eating small insects and other arthropods. The birds devote much of their time each day to catching prey, which they do in two main ways: by "hawking" for insects in flight, often high above the treetops or over water; or by gleaning them from foliage, especially the underside of leaves.

Hermits habitually visit spiders' webs, feeding on both small spiders and the insects that have been caught in the webs.

Energetics

Among birds, most hummingbirds are extremely small. Over a hundred species weigh 4.2 g or less, making them smaller than the Short-tailed Pygmy Tyrant (*Myiornis ecaudatus*), the smallest bird in any other family. Even the Giant Hummingbird of the South American Andes, which is nearly twice the size of any other hummingbird, weighs only 20 g, lighter than a House Sparrow (*Passer domesticus*). The next biggest hummingbirds weigh not much more than 12 g, or about the same as a chickadee. They include the Sword-billed Hummingbird, the Great Sapphirewing,

LEFT & OPPOSITE A maie Magenta-throated Woodstar (left) defecating while feeding at poinsettia. Nectar is processed within a few minutes leaving little more than drops of clear water. A White-tipped Sicklebill (opposite) at *Heliconia*. Note the long tongue protruding from its bill.

sicklebills, and sabrewings. The Cuban Bee Hummingbird weighs only 1.6 g and has the distinction of being the smallest bird in the world, though several other species are only marginally bigger, including several woodstars and the smaller hermits.

The amazing flying ability of the smaller hummingbirds comes at a high cost, for hovering flight on wings beating 50 to 80 times per second expends an enormous amount of energy. This is on top of an energy expenditure that is already high because of the birds' small size and high body temperature of around 102 to 108 °F (39 to 42 °C)—small, warm bodies lose heat faster than bigger, cool ones.

In fact, tiny hummers like the Bee Hummingbird and the smaller woodstars have the highest metabolic rate of any warm-

blooded vertebrate, rivaled only by the equally tiny Kitti's Hog-nosed Bat (*Craseonycteris thonglongyai*) and Pygmy Shrew (*Microsorex hoyi*). To sustain their high metabolic rate, hummingbirds must eat an enormous amount, sometimes consuming several times their body weight in nectar a day. A reliable supply of nectar, ingested every few minutes, is crucial, and it must be digested quickly to make room for more. Nectar, which contains water, highly calorific natural sugars, and not much else, is processed by the birds within 15 to 20 minutes, the waste products emerging as little more than a few crystal-clear droplets (see photo, opposite).

To conserve energy while sleeping and to cope with low temperatures and long nights, hummingbirds can enter a state of torpor, when their body temperature drops to near that of the surrounding air. They do not necessarily go into torpor every night—it depends on weather conditions and the energy reserves of the individual hummingbird. Torpor is particularly important for small species that live at high altitudes, where frosts and extremely cold nights are the norm. For tiny Scintillant Hummingbirds, weighing only 2 g, some of which live at well over 6,000 feet (1,850 m), it is a life-saving ability.

The hummingbird's tongue
Hummingbirds have an extendible tongue, up to twice as long as the bill, with a forked, fringed tip. Early naturalists, finding only insects in the birds' stomachs, thought that hummingbirds caught them in flowers, spearing or entangling them with the tip of their tongue. An alternative

theory supposed that the tongue was sticky, like that of an anteater, and that insects stuck to it when it was inserted into flowers. Later, once it was realized that nectar was the real objective, it was thought that the tongue functioned like a drinking straw, much like a butterfly's proboscis.

The way a hummingbird's tongue really works has remained a source of controversy for many years. In the 1970s it was determined that hummingbirds lap up nectar rather than suck it through a tube. At that time it was shown that the end of the tongue has two longitudinal channels and it was thought that nectar flowed into the channels by capillary action. Then, after being carried into the bill,

the nectar was squeezed out of the tongue. Recently, however, high-speed photography has revealed that the longitudinal channels open and close, trapping a volume of nectar. In real time, the tongue flicks in and out of the flower up to a dozen times a second, supplying the hummingbird with a steady stream of nectar.

Feeding strategies

Twenty or more different hummingbird species often occur together, each of them occupying a slightly different niche—the habitat that supplies everything for a species' survival. The ways in which they are segregated depend partly on their structure, especially that of their bill, and partly on their behavior.

It is often said that the length and curvature of the bills of hummingbirds fit the flowers on which they feed. This is true only in a general way and there is seldom a simple one-to-one relationship. On the contrary, most hummingbird flowers are visited by many different species. Based on their bills, hummingbirds are best divided into only two major groups, one consisting of hummingbirds with a long, often curved bill, and a much bigger group with a shorter, straight bill.

All the hummingbirds in the first group, which includes the hermits, lancebills, incas, Sword-billed Hummingbird, and others, visit specialized flowers with a long tubular corolla and abundant nectar. There is an enormous

overlap in the selection of flowers that they visit. In the Costa Rican lowlands, for example, any of the long curved flowers of *Heliconia pogonantha* or *H. longa* are likely to be visited by all the hermits in the area.

The only hermits that depend on a limited selection of flowers are the two sicklebills. Their extraordinary bills match the strongly curved corolla of *Centropogon* flowers and a few species of *Heliconia*, but are too acutely curved to be inserted easily into other flowers. Even though the sicklebills are locked into this relationship, the reverse is not true: Other hermits visit all the flowers used by sicklebills. Their bills may not fit the flowers so perfectly, but their long, flexible tongues easily compensate.

OPPOSITE & RIGHT Many hummingbirds occasionally pierce the base of large flowers to steal nectar. A few specialize in this behavior, including the Wedge-billed Hummingbird, here stealing from a Chinese Lantern (opposite), and the Stripe-tailed Hummingbird from *Poikilocanthus* (right).

The second group of hummingbirds, which have relatively short bills, are able to feed at virtually any flower with a short corolla. A clump of such flowers may attract six or seven short-billed species on most mornings and many more over a period of days or weeks. Even so, there is a tendency for the hummingbirds with the shortest bills to visit the shortest flowers, and those with longer bills to use longer flowers.

Hummingbirds can also be classified according to the ways in which they visit and exploit flowers—in other words, according to their ecological roles or "professions." The category of "high-reward trapliners" corresponds to the long-billed hummingbirds described above, which fly long distances traplining specialized flowers with a long corolla. In contrast, "territorialists" are dominant, aggressive species that defend large patches of flowers against all other hummingbirds, including members of their own species. Good examples of this type include the Rufous-tailed Hummingbird and other species of *Amazilia*, coronets, sunangels, woodnymphs, and others.

In addition to trapliners and territorialists, there are also such antisocial professions as marauders, filchers, and piercers, all of which engage in stealing nectar. Marauders and filchers steal from territorial hummingbirds, either by barging or sneaking into their territories. Piercers extract nectar through holes made in bases of flowers and play no part in pollination.

These ecological roles are useful shorthand for hummingbird behavior, but they are not rigid categories. A typical territorialist does not always defend a territory, and a trapliner defends flowers occasionally. The way a hummingbird forages is determined by opportunity and necessity, and it adjusts its behavior accordingly, from season to season, day to day, or even hour to hour. A hummingbird may be a territorialist in the morning, when nectar is flowing strongly in the flowers it is defending, then trapline intermittently when the flow decreases. A filcher that steals from a territorialist in the morning may even become a territorialist later in the day when it defends the same flowers against other small hummers. Many species are occasional piercers.

Hummingbirds & pollination

Most tropical plants rely on animals to transfer pollen from flower to flower. The most important animal pollinators are bees, but there are many others, including other insects, bats, and hummingbirds. Since flowers "want" to be pollinated, they have features that match the physical and sensory abilities of pollinators. To achieve pollination, flowers provide a reward, advertise it, and are constructed so that visitors come into contact with their stamens and stigma. The commonest reward is nectar, but others include pollen and less usual substances, such as waxes, oils, and perfumes (the latter is used by orchid bees as sexual attractants).

Nectar varies in the sugars it contains. Flowers pollinated by hummingbirds, butterflies, hawkmoths, and many bees secrete nectar rich

OPPOSITE A Rufous-tailed Hummingbird visiting Sanchezia (top) and a Green Violetear at a bromeliad (bottom), both of them probably pollinating the flowers. Note the pollen on the Violetear's crown.

in sucrose, whereas those pollinated by bats and by passerine birds (including American orioles and tanagers, Australian honeyeaters, and Old World sunbirds) have nectar rich in glucose and fructose, sugars that are also found in fruits. The significance of these intriguing differences is unknown, but they do not seem to matter much to hummingbirds and bats. Hummingbirds happily feed on leftover nectar in bat flowers, while bats routinely empty hummingbird feeders full of sucrose solution.

Flowers have evolved a variety of adaptations to attract specific pollinators. Certain plant families are particularly important for hummingbirds, some of the most notable being the Heliconiaceae, Bromeliaceae, Ericaceae, Rubiaceae, Acanthaceae, and Gesneriaceae. These flowers are ornithophilous, or bird-loving, and exhibit clear adaptations for pollination by hummingbirds. Most of them are either red, or have red bracts or leaves that advertise their presence, red being a color that is conspicuous to hummingbirds but not to most insect competitors. In addition, the flowers are diurnal, have a tubelike corolla that fits the slender bill of a hummingbird, and lack any scent (which might attract insects), and most also lack a landing platform that would provide easy access for insects. Flowers aimed at other pollinators have other characteristics. Those pollinated by hawkmoths, for example, are nocturnal, white (to show up in the dark), fragrant, and also tubular (to fit a moth's proboscis). Bat flowers are nocturnal and usually pale, with a strong musky odor, and some have "sonar guides," which bats detect by echolocation.

There are two main types of hummingbird flowers. One sort has long tubular flowers (mostly 30 to 40 mm long) that secrete copious nectar; these tend to be scattered and are usually visited by traplining hummingbirds. The other sort has short tubular flowers (mostly less than 20 mm long); these contain less nectar, but are massed together in numbers great enough to be worth defending by territorial hummingbirds. These differences have consequences for the plants. Trapliners tend to carry pollen from plant to plant, which results in cross-pollination and enhanced reproductive success. In contrast, territorial hummingbirds foster self-pollination. Sometimes a hummingbird visits so many flowers on the same plant that its face gets covered with white or golden pollen, making it look like a different species. Hummingbirds that intrude into territories to steal nectar are probably more useful to a plant than the territorial owner. Since these filchers visit flowers only briefly before being chased off, they perhaps deliver pollen to plants of the same species some distance away.

Another point to be considered is why many flowers use hummingbirds as pollinators rather than insects. After all, insects can be attracted with a smaller reward of nectar. The probable reason is that hummingbirds are more reliable as pollinators when the weather is bad, particularly at high altitudes. Bees and butterflies remain inactive when it is very wet or too cold, so flowers dependent on them fail to be pollinated. Hummingbirds are active in all kinds of weather, so it is no surprise that there are many more hummingbird-pollinated plants in the highlands than the lowlands.

Courtship, seasonality & nesting

The annual cycle of all birds includes activities that demand time, energy, and other resources beyond those required for daily maintenance. Breeding is the most critical. It requires enormously increased energy expenditure for courtship, nest-building, egg production, and feeding young.

Molting is less challenging, but still involves an increase in metabolic rate of about 30 percent. Molt immediately follows breeding, and both activities take place when food and other resources are plentiful. Birds that migrate long distances face further demands in order to accumulate migratory fat. This is a major factor for a few hummingbirds that migrate hundreds or thousands of miles. Many others undergo seasonal movements that are relatively short.

Hummingbirds are polygamous, and all nesting activities are carried out by the female unaided. Flowers and nectar are so abundant at some times of year that females can feed themselves and rear two nestlings with ease. As far as males are concerned, breeding begins and ends with courtship and mating, and many spend an inordinate amount of time advertising their services. How males attract females mostly depends on their ability, or lack of it, to defend a territory containing a rich source of nectar. Males that are dominant territorialists establish their desirability as a mate by the quality of the flowers in their territory—the better the flowers, the more likely they are to be chosen by females.

RIGHT A female Band-tailed Barbthroat building its nest attached to a palm frond. Much use is made of strands of spider silk to anchor the nest securely.

Male Fiery-throated Hummingbirds are known to defend more flowers than they need and they allow females to use the surplus.

In many hummingbird species, males establish their status by direct competition. Some join singing assemblies (leks), which range in size from as few as two or three birds to 20 or even more in the case of the Long-tailed Hermit. Most lekking males have squeaky, monotonous songs that they repeat interminably for much of the year. Within an assembly, males defend several perches and compete for central positions because females prefer to mate with the central males. For the most part, hummingbird species that perform in singing assemblies are nonterritorial for one reason or another. Hermits, for example, trapline widely dispersed flowers that are not defendable, while many small hummers are too subordinate to defend patches of rich flowers.

Several of the smallest hummingbirds advertise in a very different way. They include several North American species, Volcano and Scintillant hummingbirds in Central America, woodstars, and probably the thorntails and coquettes. They tend to live in open country or high in the canopy and have spectacular aerial displays, often accompanied by loud sound effects made by the wings.

The breeding seasonality of hummingbirds is very variable, since it depends on the abundance of preferred flowers. In the temperate areas of North America and southern South America, breeding obviously takes place in the summer of each hemisphere. In the dry forests of Central America, which stretch along the Pacific coast from Mexico to northwest Costa Rica, hummingbirds nest mainly in the dry season, when they exploit abundant flowers that are really intended to be pollinated by bees or other insects. In the mountains of Central America and in the Andes, cloud-forest hummingbirds often breed when the weather is unpleasantly cold and wet. It is the time when colorful epiphytic heaths (Ericaceae) and other favored flowers are blooming most profusely. In lowland rainforest many hummingbirds breed almost throughout the year.

Hermits' nests are different from those of typical hummingbirds and very distinctive. They are almost invariably attached to the underside of the tip of a drooping palm frond or *Heliconia* leaf, low in the forest undergrowth. The supporting leaf, arching overhead, provides a waterproof shelter and partial protection from the prying eyes of predators. When she begins to construct the nest, the female hermit has to work on the wing, using cobwebs to attach fragments of plant material to the drooping leaf. Every so often she circles the leaf, facing inward, delicately winding strands of spider silk round and round the growing nest, binding it securely to its flimsy support. The finished nest is shaped like an inverted cone and usually has a long, straggling tail of dead leaves. When sitting on the nest incubating her eggs or brooding small chicks, the female hermit always faces inward, her head and neck forced backwards into what appears to be an excruciatingly uncomfortable position. It is also noteworthy that she hovers to feed her chicks, rather than perching on the rim of the nest (see opposite).

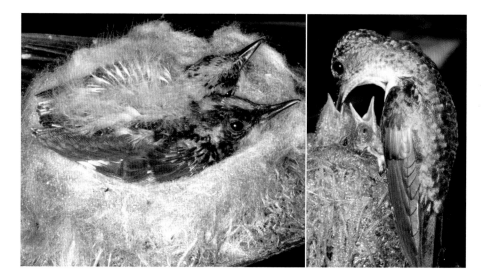

The nests of most typical hummingbirds (excepting a few South American species that construct domed nests or nest in caves) are rather similar in construction and appearance, and are easily recognized. Each tiny structure is a dainty cup saddled on a twig or branch, created from green moss, plant fibers, fern scales, downy seeds, and other soft materials, all bound together with cobwebs. Nests are often characteristically ornamented. Hummingbirds that build in the forest understory usually drape their nests with strands of moss, while those that nest in sunny positions often use lichens. The end result is always a well-camouflaged structure that is both exquisite and practical.

All hummingbirds lay two tiny, elongated white eggs that are incubated by the female for 14 to 19 days. At most hummingbird nests that have been observed, the female has returned as

ABOVE Nestling White-necked Jacobins (left) and a female Stripe-tailed Hummingbird feeding her nestlings (right). Female hummingbirds feed their young by regurgitation, delivering a mixture of nectar and tiny insects.

regularly as clockwork to feed her nestlings, every 25 to 30 minutes or so, bringing them a mixture of nectar and tiny insects. She feeds them by regurgitation, thrusting her rapier-like bill deep into their throats—an unnerving sight. The chicks stay in the nest for around 18 to 28 days, depending on the prevailing weather and availability of flowers. By the time they are ready to leave, the chicks completely fill the nest.

Sadly, many nestlings never fledge, for hummingbirds' nests fail just as frequently as those of other tropical birds, plundered by toucans, jays, squirrels, snakes, and a host of other predators.

Molt & feather care

Feathers deteriorate with age, so all birds have to replace their plumage every year to ensure that it functions efficiently. Molting requires additional energy and nutrients, and at the same time reduces a bird's performance, making it a costly event in the annual cycle of birds. In most birds, including hummingbirds, molting immediately follows breeding, the two activities being squeezed into a time of year when food is usually plentiful.

Hummingbirds undergo one complete molt a year, lasting four to five months. Flight and contour feathers molt at the same time, the renewal of the all-important primary feathers usually spanning the whole period. Primary molt begins with the short innermost primaries and proceeds outward. Hummingbirds are unique among birds in reversing the sequence of molt of their two outermost primaries, an adaptation that preserves the aerodynamic efficiency of the wingtip.

Molt is also the process through which juvenile birds attain adult plumage. Most young hummingbirds are duller versions of the adults, often with buff fringes to the feathers covering their upperparts. Full adult plumage is attained within a few weeks of fledging. The adult males of a few species, notably some of the woodstars, also molt after breeding, losing their iridescent gorget and ending up with a duller so-called "eclipse" plumage.

Hummingbirds spend a lot of time each day—as much as 70 to 80 percent according to most estimates—sitting around digesting food between feeding bouts. They devote much of this time to preening, scratching, and oiling

their plumage, keeping it in top-notch condition. Feather maintenance is essential for all birds for efficient insulation and flight, but especially so for such energetic fliers as hummingbirds. Most feather maintenance is done by preening with the bill, although long-billed hummingbirds have a hard time reaching all parts of their body that need attention. Instead, they groom the hard-to-reach places with their feet by scratch-preening. The Sword-billed Hummingbird of the South American Andes has a bill so long that it can reach hardly any of its body. To compensate, it has unusually flexible legs and feet, enabling it to scratch even the center of its back, a feat impossible for most hummingbirds. After bouts of preening, hummingbirds often body-shake to fluff and ruffle their plumage, helping it settle back into place.

Hummingbirds are fond of bathing and are often seen dipping into streams and small waterfalls, especially in the late afternoon or at dusk. Sometimes the smaller species bathe in the treetops, taking advantage of wet foliage or the small pools of water that collect in knotholes or the rosettes of bromeliads. Many hummingbirds enjoy bathing in the rain, ruffling their plumage and vibrating their wings ecstatically while spraying droplets of water in all directions.

BELOW Feather care is extremely important for hummingbirds. The birds seen preening here (left to right) are Chestnut-breasted Coronet, male Magenta-throated Woodstar, and male Violet-headed Hummingbird, while the hummingbird bathing in the rain is a male Purple-throated Mountain-gem.

Migration

Most North American hummingbirds migrate south to Mexico and Central America to escape the winter, though increasing numbers of several species now winter in Florida and along the Gulf Coast, where they depend on hummingbird feeders. The Rufous Hummingbird makes the longest migration— almost 4,000 miles (6,450 km) from Alaska to Mexico. The Ruby-throated Hummingbird is the only hummingbird to make a long migration over water—in the spring many cross the Gulf of Mexico from Central America to return to the eastern United States, a journey of up to 2,500 miles (4,000 km). Many individuals navigate around the Gulf, staying over land, while others fly nonstop for 500 miles (800 km) straight across it, a notable achievement for so small a bird.

Shorter movements are commonplace among tropical hummingbirds. In Costa Rica, for example, hummingbirds migrate in response to different flowering regimes, particularly when they finish breeding. Altitudinal movements are common as hummers shift up and down mountainsides in response to seasonal and local variations in the abundance of flowers. On the Caribbean slopes of the Monteverde Cloud Forest Reserve, the leguminous tree *Inga oerstediana* flowers profusely in May and June and is a magnet for numerous species. Brown Violetears and White-necked Jacobins move up from lower altitudes, while Coppery-headed Emeralds and Green-crowned Brilliants move down from higher elevations, and some Steely-vented Hummingbirds even cross the mountain divide from the Pacific slope.

ABOVE A male Ruby-throated Hummingbird wintering in the dry forests of Guanacaste, Costa Rica.

Two other species are present in the Monteverde region for only a few months of the year. The Magenta-throated Woodstar is a breeding visitor, with males arriving in September, a few weeks ahead of the females, to set up territories. Breeding finishes by the end of March and both sexes depart in April or May. The second species, the Scintillant Hummingbird, is a non-breeding visitor to Monteverde. Having bred on the Pacific slopes of Costa Rica's central volcanoes and highlands, Scintillant Hummingbirds spread out, moving northwest along the central divide and up and down the mountain slopes. A few birds reach Monteverde, where they remain from about March until June.

Short-distance altitudinal migrants are also commonplace in South America, particularly in the Andes, though little is known in detail about which species move or the timing and direction of their movements. In Ecuador, many hummingbirds move during the drier half of the year, from about May or June until October or November. In the Tandayapa Valley, for example, there is a changeover of species at this time, with some departing and others arriving. Buff-tailed Coronets, Purple-bibbed Whitetips, Booted Racket-tails, and Violet-tailed Sylphs become scarce and presumably move to more humid regions. Other species, including the Green and Sparkling violetears and Green-tailed Trainbearer, are dry season visitors, descending to escape the even drier conditions prevailing in the high-altitude Andean valleys. There is also an influx of Purple-throated Woodstars, including non-breeding males in eclipse plumage and many juveniles. The endangered Esmeraldas Woodstar is the most surprising altitudinal migrant to turn up in the Tandayapa Valley. It is found in its tiny breeding range, on the slopes of the coastal cordillera in Manabí and Santa Elena provinces, only during the November to April rainy season. Its non-breeding range is essentially unknown, but there have been three recent records of immature males in the Tandayapa Valley, more than 3,000 ft (900 m) higher than any other records. It seems probable that the Esmeraldas Woodstar escapes the Pacific dry season by migrating up the west slope of the Andes.

How to use this book

Species order: Most species in this book are described with an accompanying photograph. Those that are not, generally the rarer and more infrequently seen species, are in a section toward the back of the book. In both sections, species are placed in taxonomic order following *The Howard & Moore Complete Checklist of the Birds of the World* (2013). This book includes the most up-to-date checklist of all hummingbirds.

Measurements: These have been obtained from a variety of sources and are the best available. Weights have been shown in grams only because of the very small size of most hummingbirds and the difficulty of showing such small weights meaningfully in ounces. Lengths are from the tip of the bill to the tip of the tail.

Maps: The maps have been supplied by BirdLife International, the world's largest nature conservation partnership. Areas marked in red indicate where a species can be seen year-round and birds are resident. Green areas show where a species is only present during the summer breeding season, and blue areas show where a species is only present in the winter.

Status: This follows the recommendations of the International Union for Conservation of Nature (IUCN) Red List 2013.1. The categories used are: LEAST CONCERN • NEAR THREATENED • VULNERABLE • ENDANGERED • CRITICALLY ENDANGERED • EXTINCT • DATA DEFICIENT • NOT EVALUATED A full explanation of the categories can be found at **www.iucnredlist.org**

Hummingbird Directory

Crimson Topaz

MALE

Its large size, incandescent color, and aggressive behavior earned the Crimson Topaz the name "King Hummingbird" among early naturalists. There are three subspecies, varying in plumage coloration and range. Up to 20 males may gather in assemblies known as leks, advertising their availability to potential mates with complex, chattering songs, fanning displays of their wings and tails, and short, looping flights through the forest canopy. Males also defend feeding territories around flowering trees and take nectar from vines, bromeliads, and the cone-shaped scarlet inflorescences of Spiral Ginger (*Costus scaber*). The females build nests of soft plant fibers and spider silk attached to twigs and vines overhanging flowing forest streams. Though territorial around their nest sites, several females may nest in close proximity along the same stream. Males typically play no parental role, but early accounts of Crimson Topaz nesting behavior reported that males assist in nest defense.

DISTRIBUTION Subsp. *pella* occurs in southern Venezuela, Guyana, Surinam, and northern Brazil; subsp. *microrhyncha* occurs in northeast Brazil; subsp. *smaragdulus* occurs in north-central Brazil and French Guiana

HABITAT Canopy, edges, and blackwater stream corridors in lowland rainforest; 650–1,650 ft (200–500 m)

SIZE Length: 5⅛–9 in (13–23 cm). Weight: 9–18 g

STATUS Least Concern

Fiery Topaz

Often considered a subspecies of the Crimson Topaz, the Fiery Topaz lives in similar habitats in the upper reaches of the Amazon River basin. Though the ranges of the two species do not appear to overlap, the female Fiery Topaz can be distinguished from the female Crimson by black leg tufts and less extensive rufous in the outer tail feathers. Relatively little is known about how the behavior and ecology of this topaz differ from those of its sister species. Though populations of Fiery Topaz are still sufficiently large and widespread that the species is not considered threatened, its habitats are disappearing as primary rainforest is cleared for timber, agriculture, mining, and petroleum development. DNA analysis suggests that the topazes represent an early offshoot of the hummingbird family and may be equally close relatives to both the hermits and typical hummingbirds.

DISTRIBUTION Subsp. *pyra* occurs in southern Colombia, northwestern Brazil, and southern Venezuela; subsp. *amaruni* occurs in northern Peru and eastern Ecuador

HABITAT Canopy and edges of rainforest, open woodland along streams, savanna edges; 0–1,000 ft (0–300 m)

SIZE Length: 5⅞–7½ in (15–19 cm). Weight: 10.5–17 g

STATUS Least Concern

MALE

White-necked Jacobin

The male of this smart hummingbird has
a dark, shining blue head and breast, and a
metallic green upperside. The belly, tail sides,
tail tip, and base of the nape are white. The
female is usually green above and white below,
with grayish scaling on the throat and breast,
but females occasionally have male-like
plumage. Subsp. *flabellifera* is larger than the
nominate. The species usually feeds alone, but
assemblages may form around flowering trees.
It mainly forages high in the forest strata and
can be seen hovering over clearings as it actively
hawks for insects, which make up a substantial
proportion of its diet. Courting males display
to females by fanning their tails to show off
the white edges, while swooping and
diving. The nest is a shallow, soft cup
built on the upper surface of a large
leaf. This is a very widespread and
quite common bird.

FEMALE

MALE

DISTRIBUTION Subsp. *mellivora* ranges from south Mexico through Central America to Panama, Colombia, west Ecuador, eastern Peru, north Bolivia, Venezuela, the Guianas, Trinidad, and much of Brazil; subsp. *flabellifera* occurs only on Tobago

HABITAT Humid forest, mature secondary growth, forest edges and clearings, plantations; 0–2,950 ft (0–900 m)

SIZE Length: 4⅜–4¾ in (11–12 cm). Weight: 6.5–7.5 g

STATUS Least Concern

MALE

Florisuga fusca

Black Jacobin

Both sexes of this very distinctive dapper hummingbird are black with a mostly white tail and a narrow white 'U' around the flanks and lower belly. It is mainly a solitary feeder and visits a wide range of both native and introduced flower species. Rich nectar sources may attract large numbers—for example, more than 50 Black Jacobins have been seen at flowering Pink Ball Trees (*Dombeya wallichii*). It is also an avid insect-eater, mainly hawking for flies. The male courts the female with a zigzagging display flight, fanning out its tail and giving a more drawn-out version of the thin, squeaky flight call. The female builds a cup-shaped nest on a slim branch or in the base of a large leaf. Incubation and fledging periods are 16 to 17 days and 22 to 25 days, respectively. This is a fairly common bird in most of its range.

DISTRIBUTION Eastern Brazil, south through Uruguay and Paraguay into northern Argentina

HABITAT Woodland, including primary and degraded forest, well-vegetated gardens and plantations; 0–4,600 ft (0–1,400 m)

SIZE Length: 4¾–5⅛ in (12–13 cm). Weight: 7–9 g

STATUS Least Concern

White-tipped Sicklebill

Along with the Buff-tailed Sicklebill, the three subspecies of the White-tipped Sicklebill boast one of the most extreme feeding adaptations of any hummingbird. The dramatic curvature of their bills allows them to tap nectar reserves within deeply curved flowers such as *Heliconia* and *Centropogon* that are out of reach of most other species of hummingbirds. White-tipped Sicklebills perch to feed more often than most hummingbirds, probably as an aid to maneuvering their bills into the narrow openings of flowers. These quiet, unobtrusive birds are most often observed traplining from one flower patch to another through the forest understory, hover-gleaning insects in sunlit gaps, or stealing prey from spider webs. Females construct nests of tendrils and rootlets beneath the tip of a large leaf, sheltered from rain and concealed from larger birds, monkeys, snakes, and other predators. The nest is so loosely woven that the eggs may be visible through the nest wall.

DISTRIBUTION Subsp. *aquila* occurs in the eastern Andes of Colombia and northern Peru; subsp. *heterurus* occurs in the western Andes of southwestern Colombia and western Ecuador; subsp. *salvini* ranges from eastern Costa Rica to western Colombia

HABITAT Understory of humid forest, secondary-growth woodland, forest edges, riversides, typically in proximity to patches of *Heliconia*; 0–6,900 ft (0–2,100 m)

SIZE Length: 4¾–5½ in (12–14 cm). Weight: 8–12.5 g

STATUS Least Concern

MALE

Buff-tailed Sicklebill

As its name implies, the most distinctive characteristics of the Buff-tipped Sicklebill are its extreme bill shape and pale cinnamon-buff outer tail feathers with white tips. The sexes are very similar, but young birds can easily be distinguished from adults by pale feather edges on the wings and smaller patches of blue iridescence at the nape. The two subspecies are also very similar, but subsp. *gracilis* has a less heavily streaked belly and shorter bill than the nominate. Sicklebills have relatively short wings for such large hummingbirds, which reduces their flight efficiency and explains their tendency to perch rather than hover when feeding at flowers. Their long tails bob almost constantly, even at rest. In the narrow zone of overlap between the two sicklebill species, the Buff-tailed seems to prefer more open and disturbed habitats than the White-tipped Sicklebill. Its nest is similar to that of the White-tipped, comprising a loosely woven cup secured with spider silk to the underside of a large leaf.

DISTRIBUTION Subsp. *condamini* occurs in the eastern Andes of southeastern Colombia, Ecuador, and northern Peru; subsp. *gracilis* occurs in the eastern Andes of Peru and northwestern Bolivia

HABITAT Undergrowth, wooded ravines, and stream sides in humid forest, swampy forest, bamboo groves, edges, overgrown clearings, plantations; 600–10,850 ft (180–3,300 m)

SIZE Length: 5⅛–5⅞ in (13–15 cm). Weight: 8–12.5 g

STATUS Least Concern

FEMALE

Saw-billed Hermit

This is a large, robust hermit with a long and somewhat heavy-looking, straight bill that has serrated edges to the mandibles, which can be visible at close range. The male's bill-tip also has a small but distinct hook. The sexes are otherwise similar in appearance, with shiny olive-green upperparts and pale underparts marked with heavy, dark streaking; the lower cheeks and tail edges are orange. The species is a solitary trapline feeder but an individual will aggressively challenge and attempt to chase off other hummingbirds (including other Saw-billed Hermits) if it encounters them on its route. It visits flowers with long corollas, and picks insects from foliage. The nest is a typical conical structure fixed to the tip of a dangling leaf. This species is classed as Near Threatened because it is believed to be declining rather rapidly, based on the rate of deforestation in its range in Brazil.

DISTRIBUTION Southeast Brazil

HABITAT Humid forest; 0–1,650 ft (0–500 m)

SIZE Length: 5½–6¼ in (14–16 cm). Weight: 5.5–8.5 g

STATUS Near Threatened

MALE

Bronzy Hermit

The Bronzy Hermit is slightly smaller than the closely related and very similar Rufous-breasted Hermit, with which it is sometimes grouped as a single species. Both sexes are bronze green with a long, downcurved beak (slightly more downcurved in the female), darker cap and face mask, and lighter eye and mustachial stripes. The tail feathers are bronze, with black bands and white tips. The upper mandible of the bill is serrated in the male but not the female. The species feeds in the understory on insects and small spiders, and on the nectar of various flowers, especially *Heliconia* species. This bird is typically a trapline feeder. It builds a cone-like nest from plant fibers and moss, hung by a plant thread or spider's silk from an overhanging branch, leaf, or man-made construction, usually 3 to 6 feet (1 to 2 m) above ground. The species has a small range but its population appears stable.

DISTRIBUTION Eastern Honduras, eastern Nicaragua, Costa Rica, western Panama, western Colombia, western Ecuador

HABITAT Dense humid forest and forest edges, shrubland, swamps, undergrowth bordering streams; 2,950–6,550 ft (900–2,000 m)

SIZE Length: 3⅜–3⅞ in (8.5–10 cm). Weight: 3–6.5 g

STATUS Least Concern

MALE

Rufous-breasted Hermit

Also known as the Hairy Hermit, this hummingbird is closely related to the similar but slightly smaller Bronzy Hermit. Both sexes are dark green above and a reddish brown underneath, and have a long, downcurved bill (more downcurved and shorter in the female). The bird has a dark cap and face mask, with lighter eye and mustachial stripes. The female's throat and breast are lighter in color. The central tail feathers are green and the outer tail feathers reddish, both with a dark band toward the end and white tips. Both mandibles of the male's bill are serrated, whereas only the upper mandible has serrations on the female. This hermit eats small spiders and insects, and the nectar of various flowers, especially *Heliconia*. It builds a cone-like nest from plant fibers and moss, connected to an overhanging branch, leaf, or manmade construction, usually 3 to 6 feet (1 to 2 m) above ground. Two subspecies have been identified: subsp. *hirsutus* on the mainland and subsp. *insularum* in the West Indies. The species is common throughout its range, but is thought to be decreasing.

DISTRIBUTION Subsp. *hirsutus* occurs in Panama and western Colombia, through Venezuela and the Guianas, Brazil, and northern Bolivia; subsp. *insularum* occurs in Grenada, Trinidad, and Tobago

HABITAT Dense humid forest and forest edges, shrubland, swamps, undergrowth bordering streams; 0–3,300 ft (0–1,000 m)

SIZE Length: 3⅞–4¾ in (10–12 cm). Weight: 5–8 g

STATUS Least Concern

MALE

MALE

Threnetes ruckeri

Band-tailed Barbthroat

The Band-tailed Barbthroat takes its names from its striking black-and-white tail pattern and the dusky, bristly feathers of its chin and upper throat. The sexes are similar, though adult males have straighter bills and slightly brighter, more contrasting plumage than females. There are three subspecies, varying slightly in plumage coloration and range. The species forages for nectar by traplining, often "robbing" from long-flowered *Heliconia* species and *Calathea* by piercing the bases of the corollas. As with other barbthroats, spiders are among its favorite prey. Males display on low perches in the undergrowth, alone or in loose groups, singing shrill, squeaky songs while bobbing their tails. The nest is a deep, untidy-looking cup attached to the underside of a large leaf such as *Heliconia*, palm, or banana.

DISTRIBUTION Subsp. *ruckeri* occurs in northern and western Colombia and western Ecuador; subsp. *venezuelensis* occurs in northwestern Venezuela; subsp. *ventosus* ranges from eastern Guatemala to Panama

HABITAT Understory and edges of primary and disturbed lowland evergreen forest, older secondary growth, shrubbery, *Heliconia* thickets, banana plantations; 0–3,950 ft (0–1,200 m)

SIZE Length: 3⅞–4⅜ in (10–11 cm). Weight: 5–7 g

STATUS Least Concern

Dusky-throated Hermit

DISTRIBUTION Southeast Brazil

HABITAT Coastal to montane primary and sometimes secondary forest; 0–7,400 ft (0–2,250 m)

SIZE Length: 3⅞–4⅜ in (10–11 cm). Weight: 2.5–3.5 g

STATUS Least Concern

A small, dark, rather dull hermit, this bird has a generally gray-brown color scheme that is relieved by an ocher belly, green iridescence on the shoulders, and the typical hermit white face stripes and tail tip. There are no marked differences between the sexes and no known subspecies. It is a bird of humid forest from sea-level to high altitudes, and is common in some localities. The Dusky-throated Hermit is a typical hermit in terms of foraging behavior, lekking displays, and nest type, but very little detailed information has been gathered so far on its behavior and biology. It nests between October and February, and the incubation and fledging periods are 14 and 20 to 22 days, respectively. Deforestation within its range is ongoing, so while it is not currently considered to be threatened, its long-term future is insecure.

MALE

Phaethornis rupurumii

MALE

Streak-throated Hermit

This small, dark hermit species is very similar to the Dusky-throated Hermit but has a brownish-gray belly and shorter tail, and a more northerly distribution. Its scientific name comes from the Rupununi River in Guyana. The population found in Guyana and surrounding areas is of subsp. *rupurumii*, while the similar-looking subsp. *amazonicus* occurs separately, around the Amazon River. Future research may indicate that this geographically isolated population should be considered a full species. Males have darker, more streaky throats and darker tails than females. The species uses a wide range of habitat types and is often quite common. When lekking, males assemble in loose groups of up to ten birds and sing from low perches (sometimes barely above ground level). The two subspecies are reported to have very different song types. After mating, the female constructs the nest (from the tip of a hanging leaf, as is typical for hermits) alone.

DISTRIBUTION Subsp. *rupurumii* occurs from east Colombia through central-eastern Venezuela and western Guyana to northern Brazil; subsp. *amazonicus* occurs along the Amazon River in north-central Brazil

HABITAT Rainforest edges, various lighter forest types; 0–1,650 ft (0–500 m)

SIZE Length: 3⅞–4⅜ in (10–11 cm). Weight: 2.5–3 g

STATUS Least Concern

FEMALE

Phaethornis longuemareus

Little Hermit

Also known as Longuemare's Hermit, the Little Hermit is one of the smallest hummingbirds. Both the male and female are dark green above and orange underneath, and have a long, downcurved beak with a black upper mandible and black-tipped yellow lower mandible. The bird has a dark cap and face mask, with lighter eye and mustachial stripes. The male's throat is dark brown, whereas the female's is lighter. Both sexes have white undertail coverts and the tail feathers are brown with white tips. The Little Hermit is mainly a trapline feeder, accessing nectar by piercing the base of flowers with its bill. It obtains its protein and other nutrients by taking small spiders and insects from leaves and webs. It builds a cone-like nest from plant fibers and moss, connected to an overhanging branch or leaf, usually 3 to 6 feet (1 to 2 m) above ground. The species is fairly common throughout its range, although its current numbers and population trend are unknown.

DISTRIBUTION Trinidad, and northeastern Venezuela to French Guiana

HABITAT Mangrove forests, rainforest, forest edges, swamps, shrubland, plantations; 0–2,600 ft (0–800 m)

SIZE Length: 3½–3⅞ in (9–10 cm). Weight: 2.5–3.5 g

STATUS Least Concern

Minute Hermit

DISTRIBUTION A small area of coastal southeastern Brazil

HABITAT Primary and old secondary rainforest; 0–1,650 ft (0–500 m)

SIZE Length: 3⅛ in (8 cm). Weight: 2–2.5 g

STATUS Least Concern

The sexes are distinct in this tiny hummingbird, with males showing extensive green iridescence on the back and a dark brown, violet-tinged wash to the breast, while females are paler and more orange-toned. Compact and short-tailed with a proportionately very long bill, this distinctive species favors lowland rainforest. Like other hermits, it forages along a regular route or circuit, taking nectar along with tiny flies and other invertebrates. Males sing and display communally to attract females at lekking sites, and after mating the female builds her nest alone, a soft conical pouch anchored to the tip of a down-hanging leaf. Although this exquisite little bird is not considered threatened at present, its range is very small and comprises vulnerable primary rainforest, which if further depleted could lead to the species' status changing to Near Threatened.

FEMALE

Stripe-throated Hermit

A small hermit with bold facial stripes, this species has an orange rump patch and orange belly, shading to paler gray-buff towards the throat. It is most closely related to the Gray-chinned Hermit. Subsp. *striigularis* has a distinctly streaked throat and paler underparts than the other subspecies, while subsp. *saturatus* has the darkest and richest orange tones on its underparts. Subspp. *subrufescens* and *ignobilis* are intermediate between subspp. *striigularis* and

saturatus. This species traplines a low-level route, visiting mainly small flowers, but it has also been observed to steal nectar from larger flowers by piercing the corolla base. Males use leks for song and display throughout the year and breeding may occur in any month. The conical nest is bound with cobwebs to the tip of a hanging leaf, and the eggs are incubated by the female for 15 to 16 days. This is a fairly common species.

FEMALE

DISTRIBUTION Subsp. *striigularis* is found in north Colombia and western Venezuela; subsp. *saturatus* ranges from south Mexico to northwest Colombia; subsp. *subrufescens* occurs in west Colombia and west Ecuador; subsp. *ignobilis* is found in north Venezuela

HABITAT Humid forest, thick secondary growth, plantations, occasionally gardens; 0–5,900 ft (0–1,800 m)

SIZE Length: 3½ in (9 cm). Weight: 2–3 g

STATUS Least Concern

Gray-chinned Hermit

A small orange and greenish-brown hermit that closely resembles the Stripe-throated Hermit, this species has three separate populations, each of which comprises a distinct subspecies. The most widespread is subsp. *griseogularis*, while the other two are the larger and paler subsp. *zonura* and subsp. *porcullae*. The latter may be classified as a separate species when more information is available. There is slight sexual dimorphism, females being paler with longer tails and wings.

This hermit forages at low levels through the understory of dense, damp montane forests, patrolling a regular "beat" and visiting nectar-bearing flowers. It is quite a common species in some areas, but subsp. *zonura* has a particularly small range that may be of conservation concern. More study is needed to ascertain details of its behavior and breeding biology, although it is likely to share many traits with the very closely related Stripe-throated Hermit.

FEMALE

DISTRIBUTION Subsp. *griseogularis* occurs in mountains in Colombia through to northern Peru, and south and southeast Venezuela to north Brazil; subsp. *porcullae* occurs in the western Andes of northwest Peru; subsp. *zonura* occurs in northwest Peru

HABITAT Mainly humid high-ground forest; 1,950–6,550 ft (600–2,000 m)

SIZE Length: 3⅛–3⅞ in (8–10 cm). Weight: 2–3 g

STATUS Least Concern

Sooty-capped Hermit

A rather large hermit with a long bill and tail, this upland and mountain species is very gray-toned with just a hint of orange on the rump and green on the back, and strongly contrasting white eyebrows, cheek stripes, and tail feather tips. There are three subspecies—subsp. *augusti*, subsp. *curiosus*, and subsp. *incanescens*—which vary only slightly but are geographically isolated populations. The birds feed from flowers along a preferred route like other hermits, and may enter buildings in pursuit of small insects; they are quite often seen in the open. The nest, a cone-shaped pouch fixed to a hanging leaf, is bound in place with spider webs and may incorporate pieces of dried mud for added stability. Incubation and fledging periods are about 15 and 20 days, respectively. At least some populations are thought to undertake seasonal movements from higher to lower altitudes.

DISTRIBUTION Subsp. *augusti* occurs in the eastern Andes and Macarena mountains, Colombia, and the Coastal Range, Venezuela; subsp. *curiosus* occurs in the Santa Marta mountains, Colombia; subsp. *incanescens* occurs in various isolated mountains in west Guyana and south Venezuela

HABITAT Montane forests, both dry and humid, scrubland, plantations; usually 1,650–5,900 ft (500–1,800 m) but up to 11,500 ft (3,500 m)

SIZE Length: 5½–5⅞ in (14–15 cm). Weight: 4–6 g

STATUS Least Concern

MALE

Planalto Hermit

This medium-sized hermit has a dark green head and upperparts, an orange-red rump and underparts, and a slightly downcurved bill (shorter and less downcurved in the female). It has a dark eye stripe or mask, with broad white stripes above and below the eye. The uppertail feathers are reddish orange, the central tail feathers are long and have white edges, and the outer tail feathers have white tips. The bird is a

trapliner feeder, and also takes insects and small spiders from leaves and webs. It makes a cone-like nest from plant fibers, leaves, and moss, often connected to a vertical structure and protected by rocks or walls. In Brazil, the breeding season is April to September. The current global population is unknown although it appears to be stable, and the species is regarded as fairly common throughout its range.

DISTRIBUTION Northeastern to south-central Brazil, eastern Bolivia, Paraguay, northern Argentina

HABITAT Mostly mountainous areas, open and semi-open areas, savanna, dry forest, forest edges, shrubland, scrubland, parks, gardens; 1,000–6,550 ft (300–2,000 m)

SIZE Length: 5⅛–5⅞ in (13–15 cm). Weight: 4–6 g

STATUS Least Concern

FEMALE
WITH CHICK

Phaethornis eurynome

Scale-throated Hermit

This is a medium-sized hermit with a green and copper-brown head and upperparts, gray underparts, and a long, slightly downcurved bill (shorter and less downcurved in the female). It is named after its throat patch of dark feathers, edged with gray-brown borders. Its dark tail feathers have broad white edges and tips, and the long central tail feathers have white edges. It has a dark eye stripe or mask, and broad yellow-white stripes above and below the eye. Subsp. *paraguayensis* is noticeably smaller than the nominate. The species is a trapliner, feeding on the nectar of many different flowers, including *Heliconia*, *Costus*, and *Centropogon*; it also takes insects and small spiders from leaves and webs. The cone-like nest is built from plant fibers, leaves, and moss, and usually suspended from an overhanging leaf. It is often decorated with

DISTRIBUTION Subsp. *eurynome* occurs in southeastern Brazil; subsp. *paraguayensis* is found in eastern Paraguay and northeastern Argentina

HABITAT Lowland and mountainous areas, rainforest, wet grassland, savanna, shrubland; 350–7,850 ft (100–2,400 m)

SIZE Length: 5⅛–5⅞ in (13–15 cm). Weight: 4–6 g

STATUS Least Concern

Spiloma lichen, which can give the nesting bird's feathers a pink hue. The current global population is unknown, although it appears to be stable, and the species is regarded as fairly common throughout its range.

White-bearded Hermit

A rather pale, gray-toned hermit, this elegant long-tailed species has green iridescence on its back fading to a pale rump. Its name comes from its clean white throat and cheek stripe; it also has a white belly and tail tip. The sexes are similar, and no subspecies are recognized, but there is considerable variation between individuals, particularly in the face pattern, some showing much wider white stripes than others. The species is widespread and is also willing to make use of a variety of different habitat types, including secondary forest growth and plantations. It feeds in the typical hermit traplining manner, patrolling a circuit within its territory and checking each flowering plant in turn along its route. The conical, pouch-like nest is usually attached to the underside of the tip of a large hanging leaf but may be built on other downward-projecting objects, even on buildings.

MALE

DISTRIBUTION East of the Andes in Bolivia, western Brazil, Colombia, Ecuador, Peru, Venezuela

HABITAT Humid forest and woodland of various kinds, especially near rivers, mainly in the lowlands but has been recorded up to 3,950 ft (1,200 m)

SIZE Length: 5⅛ in (13 cm). Weight: 4–6 g

STATUS Least Concern

White-whiskered Hermit

This bright, shining blue-green hermit has a very long, nearly straight bill (the female's is a little more downcurved than the male's). The green extends across the breast, while the belly is grayish. It has the typical hermit facial pattern, though in older males the white stripes are almost absent. The species is a trapliner, visiting mainly long-corollaed flowers like *Heliconia*, and also picks insects and spiders from vegetation. It has a fast, darting flight, and regularly gives its single squeaking call as it forages. Males display at communal leks on and off year-round, singing a simple song consisting of repeated rasping notes and wagging their tails. The nest is fixed to the underside of a long leaf. The White-whiskered Hermit is a fairly common species, outnumbering other hermits in its range up to an altitude of about 4,000 feet (1,200 m); above this, it is mainly replaced by the Tawny-bellied Hermit.

DISTRIBUTION Western Colombia, Ecuador, possibly into southwest Panama

HABITAT Humid forest and secondary growth, scrub, plantations; 0–6,550 ft (0–2,000 m)

SIZE Length: 5⅛ in (13 cm). Weight: 4–7 g

STATUS Least Concern

MALE

Phaethornis guy

MALE

Green Hermit

This is a dark hermit with a rather long, somewhat downcurved bill. The plumage is almost entirely shining, dark blue-green, with blue bases to the tail feathers, and only females have the typical pale facial stripes; males have entirely dark faces. The nominate is larger and darker than the other subspecies; subsp. *coruscus* is the brightest; subsp. *emiliae* is smaller than the previous two; while subsp. *apicalis* is as small as subsp. *emiliae* but has the shortest bill. This is a typical hermit in its habits, though it feeds in the canopy more often than most. It visits a wide range of flowers, including introduced species. Males gather in leks that may contain several dozen individuals, each delivering its simple, repetitive song from within thick shrub. Males have been observed to defend females' nest sites. This is a very widespread species, but its abundance varies considerably across its range.

DISTRIBUTION Subsp. *guy* is found on Trinidad and adjacent northeast Venezuela; subsp. *coruscus* ranges from Costa Rica to northwest Colombia; subsp. *emiliae* is found across most of Colombia; subsp. *apicalis* inhabits the eastern slopes of the Andes from north Colombia through northwest Venezuela to southeast Peru

HABITAT Damp forest and forest edges, secondary growth, plantations; 0–7,200 ft (0–2,200 m)

SIZE Length: 5⅛ in (13 cm). Weight: 4–7 g

STATUS Least Concern

Tawny-bellied Hermit

This medium-sized hermit has dark green upperparts and a yellow-orange throat and underparts, and a very long, downcurved bill. It has two long, white central tail feathers, and the outer tail feathers have broad yellow-orange edges and tips. The male has an orange rump, and the female has a more downcurved bill and shorter wings. The bird has a dark eye stripe or mask, with pale stripes above and below the eye. Subsp. *columbianus* has a dark brown throat and breast, sometimes with a white stripe down the middle of its breast and belly. The Tawny-bellied Hermit is a trapline feeder, also taking insects and small spiders. Males display at leks, repeating a squeaky *tseep* call. The birds make a cone-like nest from plant fibers and leaves, which is often suspended from an overhanging leaf. The current global population and population trend are unknown, but the bird is regarded as fairly common throughout its range.

DISTRIBUTION Subsp. *syrmatophorus* occurs in the western Andes of Colombia to southwestern Ecuador; subsp. *columbianus* occurs in the eastern Andes of Colombia to northern Peru

HABITAT Mountainous areas, rainforest and forest edges; 2,300–9,850 ft (700–3,000 m)

SIZE Length: 5⅛–5⅞ in (13–15 cm). Weight: 4.5–7 g

STATUS Least Concern

MALE

MALE

Phaethornis koepckeae

Koepcke's Hermit

A medium-sized hummingbird with dark green upperparts and a brown and orange throat and underparts, Koepcke's Hermit has a long, straight bill that is shorter and very slightly downcurved in the female. It has two long, white-tipped central tail feathers, and the outer tail feathers have broad yellow-orange tips. The bird also has a dark mask, with white stripes above and below the eye. It is very closely related to the Needle-billed Hermit, differing from that species in its dark brown throat and preference for higher altitudes. It is also related to the Tawny-bellied Hermit, but doesn't share that species' white belly stripes. Koepcke's Hermits are trapline feeders and also catch insects and small spiders. The birds make a small cone-like nest from plant fibers and leaves, which is suspended from an overhanging leaf. This species was discovered in 1977 in central Peru, and has since been found only on the slopes of seven isolated mountainous areas in the country. The current total population is estimated at between 10,000 and less than 20,000 individuals, but as its habitat is threatened by growing deforestation, the species is classified as Near Threatened.

DISTRIBUTION Eastern Andes of Peru

HABITAT Mountainous areas, undergrowth in rainforest and forest edges; 1,500–4,250 ft (450–1,300 m)

SIZE Length: 4¾–5½ in (12–14 cm). Weight: 4–6.5 g

STATUS Near Threatened

Straight-billed Hermit

One of the drabber hermits, this species has an almost straight bill and is mainly gray-brown with a touch of green iridescence on the back and tail. There are two subspecies: subsp. *bourcieri*, across most of the range; and the considerably larger subsp. *major* in Brazil south of the Amazon. Two color forms of subsp. *bourcieri* occur, one with warm brown tones on the underside, the other with grayer coloration; subsp. *major* occurs only in a gray form. This is a common hummingbird of the forest understory, and has the typical hermit foraging strategy of circuit-patrolling or traplining. Flowers visited include tubular *Manettia* species, and bromeliads such as *Guzmania* and *Vriesea*. The male's insect-like song is given almost year-round, accompanied with a tail-waggling display. The long conical nest is attached to a hanging leaf tip. Incubation takes 17 to 18 days, and the chicks fledge after 23 days.

DISTRIBUTION Subsp. *bourcieri* ranges from eastern Ecuador, northern Peru, and south-eastern Colombia across to southern Venezuela, the Guianas and northern Brazil; subsp. *major* is found in northern Brazil

HABITAT Most types of densely forested habitat, especially hilly rainforest; 0–5,250 ft (0–1,600 m)

SIZE Length: 4¾–5⅛ in (12–13 cm). Weight: 4–5 g

STATUS Least Concern

MALE

Long-billed Hermit

A very long-billed and long-tailed hermit, this species (sometimes known as Western Long-billed Hermit) is closely related to the Long-tailed Hermit and the two are sometimes considered conspecific. It is a warm bronzy-brown color with unmarked pale or buff underparts and prominent pale facial stripes. The subspecies, which apart from subsp. *baroni* intergrade through the species' range, vary subtly in size, and in the tone and intensity of color and thickness of the facial stripes: subsp. *mexicanus* is the largest and darkest; subsp. *cephalus* is the brightest; and subsp. *baroni* the grayest. The Long-billed Hermit is a traplining species and follows a route about 0.5 miles (800 m) long, often moving at high speed. It is curious and may closely approach an observer, but maneuvers to stay unseen behind the observer's head. Males lek in groups, calling ceaselessly while pumping their tails. The bird is generally considered fairly common, though some subspecies are very localized.

DISTRIBUTION Subsp. *longirostris* is found in southern Mexico; subsp. *griseoventer* occurs in west Mexico; subsp. *mexicanus* occurs in southwest Mexico; subsp. *cephalus* ranges from eastern Honduras to northwest Colombia; subsp. *susurrus* is found in north Colombia; subsp. *baroni* occurs well south of the other forms, in west Ecuador and northwest Peru

HABITAT Humid forest interiors, edges, and clearings, secondary growth, other densely vegetated habitats; 0–8,200 ft (0–2,500 m)

SIZE Length: 5⅛–6¼ in (13–16 cm). Weight: 4–7.5 g

STATUS Least Concern

MALE

Long-tailed Hermit

Also known as the Eastern Long-tailed Hermit, this medium-sized hummingbird has a green-brown head and upperparts, pale gray-buff underparts, and a downcurved bill with a red-tipped lower mandible. It has two pale bands on its longest uppertail coverts. The outer tail feathers are dark with white tips, and the two long central "streamer" feathers are dark at the base, then mostly white. The bird has a dark eye stripe or mask, with white stripes above and below the eye, and a central white throat stripe. The female is slightly smaller, with a shorter, less downcurved bill and shorter wings. Subsp. *muelleri* has a darker throat and underparts, and both sexes have a less downcurved bill. Long-tailed Hermits are trapline feeders, favoring *Heliconia* and *Passiflora* flowers, and also feed on insects and small spiders. They make a cone-like nest from plant fibers and moss, which is often suspended from an overhanging leaf.

The current global population is unknown, although the species is regarded as common and numbers appear to be relatively stable throughout its range.

MALE

DISTRIBUTION Subsp. *superciliosus* is found in southern Venezuela, the Guianas, and northern Brazil; subsp. *muelleri* is found in northern Brazil

HABITAT Lowland areas, mostly undergrowth in rainforest, forest edges, undergrowth bordering streams; 0–4,900 ft (0–1,500 m)

SIZE Length: 5⅛–5⅞ in (13–15 cm). Weight: 4–6 g

STATUS Least Concern

Great-billed Hermit

A large, dark, very long-billed hermit, this
hummingbird occurs in three separate
populations over a wide swathe of South
America and is usually considered to have six
subspecies: subsp. *malaris*, subsp. *margarettae*
(sometimes considered a separate species,
Margaretta's Hermit), subsp. *insolitus*, subsp.
moorei, subsp. *ochraceiventris*, and subsp.
bolivianus. These forms vary in size and general
plumage tone, with the northerly subsp. *malaris*
being the largest and darkest of the group.
A typical hermit in its feeding and breeding
behavior, it forages through the understory by
traplining and is known to feed from long-
flowered *Heliconia* and *Pitcairnia* species, as well
as taking flies and spiders. Males assemble in
leks to sing for female attention. Although the
species is generally common, subsp. *margarettae*
of the eastern coast of Brazil is restricted to
forest remnants and is threatened by further
habitat loss.

MALE

DISTRIBUTION Three separate
populations: subsp. *malaris* is
found in the Guianas and
northeast Brazil; subsp.
margarettae occurs in coastal
eastern Brazil; the remaining
four subspecies range across
the largest area, covering parts
of Bolivia, Ecuador, Colombia,
Venezuela, Peru, and Brazil

HABITAT Dense rainforest
understory and scrub; usually
0–1,950 ft (0–600 m) but up
to 5,400 ft (1,650 m) and
occasionally higher

SIZE Length: 5⅛–6¾ in
(13–17 cm). Weight: 4–10 g

STATUS Least Concern

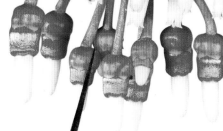

Green-fronted Lancebill

This is a fairly small, dark hummingbird with a very long, slightly upturned bill adapted for accessing long, down-hanging flowers. It is named for the glittering green patch on its forehead. Females are paler and less intensely toned than males.

There are two subspecies: subsp. *ludovicae*; and the smaller subsp. *veraguensis*, in which the sexes are more distinct. It occurs in wet montane forests, and feeds near the lower canopy, mainly from epiphyte flowers with long corollas, such as *Psammisia*. It is a trapline feeder but males sometimes defend clumps of suitable flowers; it also flycatches over water, and picks insects from vegetation. The bulky cup- or cylinder-shaped nest is often built in a shady ravine hollow on a dangling leaf or root. The incubation and fledging periods are 19 and at least 25 days, respectively; fledged youngsters are grayer and drabber than adults.

MALE

DISTRIBUTION Subsp. *ludovicae* occurs south across the Andes from east Panama, Colombia, and west Venezuela down to northwest Bolivia; subsp. *veraguensis* occurs from north-central Costa Rica to west Panama

HABITAT Cloud forest; mainly 2,450–8,550 ft (750–2,600 m), but down to 1,650 ft (500 m) or lower when not breeding

SIZE Length: 4⅜–5⅛ in (11–13 cm). Weight: 5.5–6 g

STATUS Least Concern

Blue-fronted Lancebill

This lancebill shows more obvious sexual dimorphism than the Green-fronted, the male having a very dark face and underside and a gleaming violet forehead patch. The female looks like a slightly paler and greener version of the female Green-fronted, with a blue-green forehead patch that varies considerably in size. There are two subspecies with widely separated distributions: subsp. *johannae*; and the paler, shorter-billed subsp. *guianensis*. The species uses its very slim and slightly upcurved bill to probe downward-hanging epiphyte flowers in the lower canopy of the forest—it also feeds at lower levels. Its nest is a long cylinder or cone with a cup-shaped depression on top, usually attached to a rocky overhang in a cave. Much of its preferred habitat has been lost in some areas but it seems to survive reasonably well in suboptimal habitats and is often quite common.

DISTRIBUTION Subsp. *johannae* occurs in the eastern Andes from central-eastern Colombia to northeast Peru; subsp. *guianensis* occurs from western Guyana and Venezuela into north Brazil

HABITAT Wet, hilly forest with ravines, forest edges, cocoa plantations; usually 1,300–5,250 ft (400–1,600 m)

SIZE Length: 3⅞–4⅜ in (10–11 cm). Weight: 4 g

STATUS Least Concern

FEMALE

Schistes geoffroyi

MALE

Wedge-billed Hummingbird

This is a small green-bronze hummingbird with a short, broad-based bill that narrows to a sharp point. Both sexes have a black cap and face, a white spot behind the eye, a bright green throat (duller in the female) with a white collar divided at its base, and violet-blue areas on each cheek. The rounded tail feathers are green and bronze, with white undertail coverts. Subsp. *albogularis* has a more brilliant violet-blue throat with a continuous white collar (in the female this extends to cover the throat), and green-bronze undertail coverts; and subsp. *chapmani* has less white on its throat and underparts. The bird feeds at low to medium levels, and is a specialist nectar-robber, taking nectar by piercing the base of the flower's corolla with its bill. Its small cup-like nest is constructed from seed and cactus fibers and decorated with lichen on the outside. The bird has a comparatively large range and a stable population, although numbers are yet to be quantified.

DISTRIBUTION Subsp. *geoffroyi* occurs in northern Venezuela, central Colombia, central Ecuador, and southern Peru; subsp. *albogularis* ranges from western Colombia to western Ecuador; subsp. *chapmani* occurs in central Bolivia

HABITAT Dense cloud forest and forest edges; 2,950–6,550 ft (900–2,000 m)

SIZE Length: 3⅜–3¾ in (8.5–9.5 cm). Weight: 3.5–4 g

STATUS Least Concern

Hooded Visorbearer

DISTRIBUTION Northeastern Brazil

HABITAT Mountainous, arid, and semiarid plateau areas with scrubland; 3,100–5,250 ft (950–1,600 m)

SIZE Length: 3½–3⅞ in (9–10 cm). Weight: 4–5 g

STATUS Near Threatened

The male of this small greeny-bronze hummingbird has a black cap and face, a small white spot behind the eye, a bright green throat and forehead, white bands and a brilliant red spot at the base of the throat, and bronze-red tail feathers. The female is very similar but has brown patches either side of the head instead of black. The bird feeds at low levels, usually below waist height, on the nectar of various flowers, including bromeliads and cacti, and members of the Rubiaceae and Loranthaceae families; it also takes small insects and spiders. It builds a small cup-like nest in a forked gap between branches using seeds, cactus fibers, and moss. Although the Hooded Visorbearer is considered fairly common and has a stable population, it has been classified as Near Threatened because of its small range and the loss of its habitat to cattle ranching.

MALE

Brown Violetear

Much drabber than the other violetears, this species is mainly dark brownish gray, with buffish fringes to the lower back and rump feathers that results in a scaled appearance. There is a small glistening blue-green throat patch and flared, shining, dark violet ear coverts and neck sides that are typical of the genus. Brown Violetears feed mainly in the canopy from epiphytes and flowering trees, such as *Inga* and *Clusia*, and while not usually territorial will aggressively challenge smaller hummingbirds. They hawk for flies at canopy height over clearings or lower down over water. Males form small leks (up to eight birds) in the canopy and sing to attract females. The nest is a small cup built saddle-style over a twig in a low shrub. This very widespread species is more dependent on forest than its congeners and is declining in areas where there is significant deforestation.

DISTRIBUTION Guatemala and Belize through Central America to Colombia, south through central Ecuador, east Peru and north Bolivia, east across Venezuela, Trinidad, Guyana, and south to eastern Brazil

HABITAT Forest edges and clearings, plantations, secondary growth; 350–9,200 ft (100–2,800 m)

SIZE Length: 4⅜–4¾ in (11–12 cm). Weight: 6–7 g

STATUS Least Concern

MALE

Colibri thalassinus

Green Violetear

This is a dazzling hummingbird with highly iridescent plumage. The plumage is shining greenish blue with a broad patch of deep, velvety violet across the cheeks and neck sides. The five subspecies vary principally in the coloration of the underside: subsp. *thalassinus* has a violet-blue chest patch; subsp. *cabanidis* may have a little dark blue on the chest; subsp. *cyanotus* has rather pale turquoise underparts; subsp. *kerdeli* is similar to *cyanotus* but has coppery-golden reflections on the rear crown; and subsp. *crissalis* is paler than *cyanotus* and has a buff rather than blue-green undertail. This species forages at all levels in the forest strata and often perches near the canopy. Males form small leks to attract females. The nest is built fairly low in a shrub and is a bulky, well-insulated structure. Post-breeding, most Green Violetears move to lower elevations. This is a common species and can thrive in partly deforested areas.

MALE

DISTRIBUTION Subsp. *thalassinus* is found from central and south Mexico to north-central Nicaragua; subsp. *cabanidis* occurs in Costa Rica and west Panama; subsp. *cyanotus* occupies mountains in Venezuela, Colombia, and Ecuador; subsp. *kerdeli* is found in northeast Venezuela; and subsp. *crissalis* is found in the Andes in Peru, Bolivia, and far northwest Argentina

HABITAT Forest edges, secondary growth, scrub and pastures on mountain slopes, plantations, gardens; 5,600–14,750 ft (1,700–4,500 m)

SIZE Length: 4⅛–4½ in (10.5–11.5 cm). Weight: 5–5.5 g

STATUS Least Concern

Sparkling Violetear

This is a rather large, dark, colorful violetear. The plumage is primarily green, with blue tail sides and flared blue-violet ear coverts and neck sides. Subsp. *coruscans* has some blue on the throat and belly, while subsp. *germanus* has much more extensive blue on the head and underside, and an entirely blue tail. A melanistic form with sooty gray-green and blue plumage has been recorded. The species feeds from ground level to the canopy and is aggressive and territorial, dominating all other species. The nest is built low down, on a horizontal or dangling twig, and the incubation and fledging periods are 17 to 18 days and 20 to 22 days, respectively. Birds nesting at higher elevations have very high breeding success due to low rates of predation. Sparkling Violetears move to lower elevations after breeding. The species is fairly common, even occurring in major cities.

DISTRIBUTION Subsp. *coruscans* ranges from northwest Venezuela and Colombia through Ecuador and Peru to Bolivia and northwest Argentina; subsp. *germanus* is found in south Venezuela, east Guyana, and northern Brazil

HABITAT Forest edges and clearings, open woodland, parks, well-vegetated gardens, plantations; 5,600–14,750 ft (1,700–4,500 m)

SIZE Length: 5⅛–5½ in (13–14 cm). Weight: 6.5–8.5 g

STATUS Least Concern

MALE

White-vented Violetear

This beautifully colored species has the violet cheek patch (flared in display) and neck patch typical of its genus. It is otherwise shining green, with a strong blue wash on the head, breast, and tail in the male, and a whitish-buff undertail. This is primarily an upland (but not montane) species and is most likely to be encountered at 3,300 to 4,900 feet (1,000 to 1,500 m). As with other violetears, the White-vented is aggressive and territorial, and usually prevails in its encounters with other species. It begins feeding early in the morning, so is often active before other hummingbirds. When not feeding, it perches on an exposed twig overlooking its patch. From here it keeps watch for intruders, calling constantly. It is common across its range and can be found in many protected areas as well as around towns and villages.

DISTRIBUTION Bolivia, Paraguay, much of Brazil, and into north Argentina

HABITAT Lightly vegetated habitats such as savanna, scrub, and sparse forest; 0–11,800 ft (0–3,600 m)

SIZE Length: 4¾–5⅛ in (12–13 cm). Weight: 5.5–7 g

STATUS Least Concern

MALE

Tooth-billed Hummingbird

This striking species has a very long, straight bill, which in the male has tooth-like serrations on the inner edge and a clear hook at the tip. The plumage is mainly rather grayish, with iridescent green on the back, bold blackish streaks on the underside, a black-bordered white rump band, a chestnut crown, and a blue nape. The female is generally duller than the male. This species' unusual bill morphology is thought to be an adaptation for stealing nectar (by piercing the base of corollas) from flowers that are too long to access directly, but may also be related to taking larger live prey such as spiders—males have been observed probing rolled-up leaves to extract spiders. It is a fast flyer and feeds primarily in upper levels in the forest strata. In its prime habitat this is a common species, but deforestation is causing its decline.

DISTRIBUTION East Panama and western Colombia to northwest Ecuador

HABITAT Primary forest, sometimes forest edges and secondary growth; 0–3,600 ft (0–1,100 m)

SIZE Length: 4¾–5½ in (12–14 cm). Weight: 7–9 g

STATUS Least Concern

MALE

Horned Sungem

This is a small, short-billed, long-tailed hummingbird. Both sexes are light, shining yellow-green above and white below, but the male's face and throat are black and his head is ornamented with orange-edged violet tufts. This species uses many habitat types and feeds from a variety of flowers, including *Citrus* and *Lantana*, usually at lower levels, as well as flycatching. The nest, a tiny compact cup built on a forking branch in a bush, is usually close to the ground, with its outer surfaces camouflaged by lichens. The female incubates the two eggs for 13 days, and the chicks fledge at about 21 days. The Horned Sungem is a migratory species in at least part of its range, its arrival and departure determined by the blooming time of its preferred flowers. It is a common bird, and its population is increasing.

DISTRIBUTION Primarily central and eastern Brazil to eastern Bolivia, but extends patchily north to south Suriname

HABITAT Various wooded habitats, including riverine forest, cerrado, and gardens; 0–3,300 ft (0–1,000 m)

SIZE Length: 3¾–4⅜ in (9.5–11 cm). Weight: 2–2.5 g

STATUS Least Concern

MALE

Purple-crowned Fairy

Green above and white below with a short bill, this elegant medium-sized hummingbird shows interesting sexual dimorphism. The male has violet cheeks and crown, while the female has black cheeks, a green crown, and a noticeably longer, more wedge-shaped tail than the male. A common hummingbird of wet lowland forest, this species feeds at all levels from ground to canopy, and almost always takes nectar from flowers with long corollas, piercing them at the base. The nest, built 13 feet (4 m) or more above ground level, is a cone formed mainly from downy plant seeds, in which the female incubates her clutch of two eggs for 16 or 17 days. The chicks are black-skinned with gray down, acquiring the full female-like juvenile plumage and fledging by about 20 days. This is a common bird throughout its range.

DISTRIBUTION From southeast Mexico through Central America to west Colombia and southwest Ecuador

HABITAT Wet forest and secondary growth; 0–3,300 ft (0–1,000 m)

SIZE Length: 4½–5⅛ in (11.5–13 cm). Weight: 5–5.5 g

STATUS Least Concern

FEMALE

Green-tailed Goldenthroat

This is the smallest and most vivid green of the three goldenthroat species. Like the other two, it has blackish-brown wings, and also has gray cheeks and a fleck of white above and below the eye. The sexual dimorphism is slight, females having subtle dark mottling on their throats. There are two subspecies: subsp. *theresiae*; and subsp. *leucorrhous*, which has a white undertail. It is a lowland species of open countryside, and like the other goldenthroats forages by moving low and inconspicuously along its traplining route, favoring flowers of the family Melastomataceae. Males sometimes defend a feeding territory around a particularly good cluster of flowers. The female builds her nest on a branch fork in a low shrub, and incubates the two eggs for 14 days. The chicks fledge at 20 to 28 days old. In suitable habitat this is a very common species.

MALE

DISTRIBUTION Subsp. *theresiae* occurs in the Guianas and north-central Brazil; subsp. *leucorrhous* occurs from east Colombia and southern Venezuela to northwest Brazil and northeast Peru

HABITAT Dry, bushy savanna and forest edges; usually 350–1,000 ft (100–300 m)

SIZE Length: 3½–3⅞ in (9–10 cm). Weight: 3.5–4 g

STATUS Least Concern

Ruby-topaz Hummingbird

A small hummingbird, this species has marked sexual dimorphism. The female is plain gray-green with a whitish underside and rufous tail sides, while the male is dark brown with a dazzling red crown, yellow breast, and orange tail. This spectacular plumage, shown off to its best advantage in the male's dancing courtship display, made the species highly sought after in the Brazilian bird trade before legal sales ended in 1970. It is very widespread and often common, with up to 21 pairs per square mile (eight pairs per square kilometer) in Trinidad. Feeding birds flycatch and visit many flower types. Males sometimes attempt to defend feeding territories, but larger hummingbirds will displace them. The female incubates two eggs in the tiny cup nest for 15 or 16 days. The youngsters, in female-like plumage, fledge at 19 to 22 days. The species is migratory, though the timing and direction of migration varies across its range.

DISTRIBUTION East Panama and most of Colombia across Venezuela to the Guianas, south through northeast and central Brazil to east Bolivia. Also on Trinidad and Tobago and other nearby islands

HABITAT Scrub, farmland, savanna, gardens; usually 0–1,650 ft (0–500 m) but non-breeding birds occur up to 5,600 ft (1,700 m)

SIZE Length: 3⅛–3½ in (8–9 cm). Weight: 4–5 g

STATUS Least Concern

MALE

Green-breasted Mango

The widespread Green-breasted Mango is the northern counterpart of the Black-throated Mango. The two appear virtually identical in the field, and there are four subspecies of the Green-breasted, complicating understanding of their distribution. The Green-breasted is the only mainland mango species north of Costa Rica, but some of its scattered southern populations overlap with those of the Black-throated. Females are similar to the female Black-throated except for blue-green iridescence in the black midline stripe. The Green-breasted prefers more open habitats and is often seen perched on bare twigs atop tall trees or hawking insects high above ground. In dry forest habitats, it depends heavily on flowering trees, including the filamentous white flowers of *Inga*, the blade-like scarlet blooms of *Erythrina*, and the pink or yellow trumpets of *Tabebuia*. As in other mango species, females typically build their nests straddling high open branches.

MALE

FEMALE

DISTRIBUTION Subsp. *prevostii* ranges from eastern and southern Mexico, through Belize and Honduras to El Salvador; subsp. *gracilirostris* occurs in Honduras, El Salvador, Nicaragua, and central Costa Rica; subsp. *hendersoni* occurs on Providencia and San Andrés islands; subsp. *viridicordatus* is found in northeastern Colombia and northern Venezuela. It is a rare visitor to the United States, including Texas, North Carolina, Georgia, and Wisconsin.

HABITAT Secondary forest, edges and clearings, savannas, plantations, pastures, mangroves; 0–4,900 ft (0–1,500 m).

SIZE Length: 4⅜–4¾ in (11–12 cm). Weight: 7 g

STATUS Least Concern

Anthracothorax veraguensis

FEMALE

Veraguas Mango

Previously considered a form of the Green-breasted Mango, the Veraguas Mango is named for the Panamanian province at the heart of its range. Both sexes and all ages are similar in appearance to the closely related Green-breasted and Black-throated mangoes, but in good light the underparts of adult males glow with intense blue to green iridescence, and they lack a black midline stripe. Veraguan Mangoes often gather in numbers at flowering trees such as *Inga* and *Erythrina*, sparring over feeding rights. When displaying, territorial males may perch on conspicuous bare twigs with their silky white flank tufts exposed atop their folded wings. Females build their nests straddling exposed branches in open tree canopies. The species appears to be expanding its range along Pacific mountain slopes in southern Costa Rica, perhaps in response to deforestation.

DISTRIBUTION Pacific slopes of western Panama, southern Costa Rica, and adjacent islands

HABITAT Open areas with shrubs and scattered trees, wooded stream edges, pastures; 0–3,950 ft (0–1,200 m)

SIZE Length: 4⅜–4¾ in (11–12 cm). Weight: 7 g

STATUS Least Concern

Black-throated Mango

DISTRIBUTION Subsp. *nigricollis* is found in Central and eastern Panama, Colombia, northeastern and western Ecuador, eastern Peru, northern Bolivia, Venezuela, the Guianas, Brazil, Trinidad and Tobago, northern Uruguay, eastern Paraguay, and extreme northeastern Argentina; subsp. *iridescens* occurs in southwestern Colombia, southwestern Ecuador, and northwestern Peru

HABITAT Forest edges, clearings, cultivated areas; 0–5,750 ft (0–1,750 m)

SIZE Length: 3⅞–4¾ in (10–12 cm). Weight: 5.5–7 g

STATUS Least Concern

The Black-throated Mango is the most widespread of the mango hummingbirds, and is found over most of northern and central South America. The deeply iridescent underparts of the adult male are split by a velvety black midline stripe that extends from the chin to the belly. Adult females and juveniles are similar to those of other mainland mango species, with boldly patterned underparts split by a solid black midline stripe, and outer tail feathers banded in wine purple, blue-black, and white. Females build lichen-covered nests on open horizontal branches. Nineteenth-century ornithologist John James Audubon painted this species from a specimen provided by amateur naturalist Dr. Benjamin B. Strobel, who claimed to have collected it in Key West, Florida.

MALE

MALE

Anthracothorax dominicus

Antillean Mango

The male Antillean Mango perfectly illustrates its genus name, which is Greek for "coal-black chest." Females are plain white below, while juvenile males have a black midline stripe like females and juveniles of typical mango species. Females of the Hispaniolan subspecies (subsp. *dominicus*) have more colorful tails than the subspecies found in Puerto Rico and the Virgin Islands (subsp. *aurulentus*). The Antillean Mango strongly prefers nectar plants with long tubular flowers, including morning glories, bromeliads (*Pitcairnia*), and flowering trees such as San Bartolome (*Cordia rickseckeri*) and Roble Cimarrón (*Tabebuia haemantha*). It is the largest of the three hummingbird species found on Hispaniola, but in Puerto Rico it competes with the similar-sized Green Mango.

DISTRIBUTION Subsp. *dominicus* is found on Hispaniola and adjacent islands; subsp. *aurulentus* occurs on Puerto Rico and the adjacent island of Culebra, and on Anegada in the British Virgin Islands

HABITAT Clearings, scrub, gardens, shade coffee plantations; 0–8,500 ft (0–2,600 m)

SIZE Length: 4⅜–4⅞ in (11–12.5 cm). Weight: 4–9 g

STATUS Least Concern

Jamaican Mango

This is one of four hummingbird species that thrill visitors to Jamaica's botanical gardens and feeding stations, though the species is common and widespread in natural habitats as well. Adult females are less iridescent versions of adult males, with narrow, pale tips on the outer tail feathers. Young males resemble females except for deep blue iridescence on the throat. Jamaican Mangoes feed extensively from the flowers of cacti such as the Cochineal Pricklypear (*Opuntia cochenillifera*). Males will defend feeding territories around flowering trees, including the Scarlet Cordia (*Cordia sebestena*), orchid trees (*Bauhinia*), and the introduced ornamental African Tulip Tree (*Spathodea campanulata*). After breeding, many individuals will move from the lowlands into middle and upper elevations, following the flowering cycles of mountain plants.

DISTRIBUTION Jamaica

HABITAT Lowland and mountain forest edges, semiarid woodlands, banana and shade coffee plantations, gardens; 0–2,600 ft (0–800 m)

SIZE Length: 4⅜–5⅛ in (11–13 cm). Weight: 9 g

STATUS Least Concern

MALE

Green-throated Carib

This hummingbird is the same length as but more lightly built than the Purple-throated Carib, and is mainly glossy green with a black belly and a little blue gloss on the rump, breast, and tail. There are two subspecies, subsp. *holosericeus* and subsp. *chlorolaemus*, the latter darker-throated with more extensive blue on the breast. It is an adaptable species, occurring in both forested and open habitats, and mainly feeds at 3 to 13 feet (1 to 4 m) above ground. It may defend large clusters of *Lantana* or *Kalanchoe* flowers. The cup nest, constructed from plant materials including bark and soft plant seeds, is built on a twig, its sides overlapping the support like a saddle. Incubation takes 17 to 19 days. The chicks fledge at about three weeks and take another three to four weeks to reach full independence. Post-breeding, birds tend to disperse to higher elevations.

DISTRIBUTION Subsp. *holosericeus* occurs in Puerto Rico and the Lesser Antilles except Grenada; subsp. *chlorolaemus* is restricted to Grenada

HABITAT Forests, rainforest, cultivated areas; mainly 0–1,650 ft (0–500 m)

SIZE Length: 4⅜–4⅞ in (11–12.5 cm). Weight: 5–8 g

STATUS Least Concern

MALE

Eulampis jugularis

FEMALE

Purple-throated Carib

Both sexes of this fairly small but stocky hummingbird are blackish with iridescent blue-green wings, rump, and tail, and a purple-red throat and breast patch. The smaller female has a longer and more downcurved bill than the male. This difference in bill shape reflects different feeding behavior—males feed preferentially from *Heliconia caribaea*, while females prefer *H. bihai*, which has a longer corolla. In areas where *H. caribaea* is rare or absent, *H. bihai* flowers occur in two different shapes (morphs), one of which is shaped like the flowers of *H. caribaea* and caters to male Purple-throated Caribs. This indicates a very close coevolutionary link between *H. bihai* and its key pollinator. Males are territorial around prime feeding sites, while females wander more widely and do not defend feeding grounds. Females are, however, fiercely defensive around their nest, a neat cup hung on a branch fork.

DISTRIBUTION Lesser Antilles, from Saba to St. Vincent

HABITAT Primary and secondary forest, edges of forest, sometimes gardens and plantations; 2,600–3,950 ft (800–1200 m)

SIZE Length: 4⅜–4¾ in (11–12 cm). Weight: 7–12 g

STATUS Least Concern

Orange-throated Sunangel

This is a fairly small hummingbird with a short, straight bill and a short but heavy, broad tail. It has shining yellow-green plumage, the male with an orange throat and breast and a small orange forehead patch, and the female with a browner throat marked with fine dark speckles and no distinct forehead patch. The species occupies high montane cloud forest, where it feeds at low levels and drives other hummingbirds away from its territory, which is based around flower clusters. It usually feeds while clinging to a nearby twig or the flowers themselves, and flicks its wings or holds them raised when about to challenge an intruder. It also joins forces with other small birds to mob owls and other predators. Its soft trilling call when feeding is distinctive and makes it easy for observers to find the bird.

DISTRIBUTION Northwest Venezuela and the eastern Andes in Colombia

HABITAT Cloud forest and nearby open and scrubby habitats; mainly 6,550–10,500 ft (2,000–3,200 m)

SIZE Length: 3⅞–4⅜ in (10–11 cm). Weight: 4–4.5 g

STATUS Least Concern

MALE

Amethyst-throated Sunangel

This species is predominantly green with a purple (yellowish in females) throat patch and narrow white breast band. There are six subspecies: subsp. *amethysticollis*; subsp. *apurimacensis*, described in 2009; subsp. *laticlavius*, with a paler throat patch and blue-green forehead patch; subsp. *decolor*, with a darker throat patch and grayer belly; subsp. *clarisse*, with a pale throat and narrow white breast band; and subsp. *violiceps*, with a pale throat and broad white breast band. The latter two subspecies are sometimes split as Longuemare's Sunangel *H. clarisse*. The species has been known to hybridize with four other hummingbird species (none of them other sunangels) in the wild. Aggressive and territorial, it defends flower clusters and also spends much time catching insects in flight. It may associate with mixed feeding flocks of other small birds. The nest is usually attached to a dangling piece of moss, below an overhanging leaf or other shelter.

MALE

DISTRIBUTION Subsp. *amethysticollis* occurs in the eastern Andes in southern Peru and northwest Bolivia; subsp. *apurimacensis* occurs in southern Peru; subsp. *laticlavius* occurs in southern Ecuador and north Peru; subsp. *decolor* occurs in the eastern Andes of central Peru; subsp. *clarisse* occurs in the eastern Andes of Colombia and west Venezuela; subsp. *violiceps* occurs in the Perijá mountains on the Colombia–Venezuela border

HABITAT Cloud forest, dwarf forest, dense and humid montane scrub; 5,900–10,500 ft (1,800–3,200 m)

SIZE Length: 3⅞ in (10 cm). Weight: 5–6 g

STATUS Least Concern

Gorgeted Sunangel

Very similar to Amethyst-throated and Merida sunangels, this is a darker bird with a dark violet throat patch and broad white crescent below; the female's throat is mottled dark green with a hint of purple. The tail, which is more deeply notched than those of the other species, has a dark violet-blue gloss. Although several color variations have been noted, the species is not considered to have any subspecies. This bird has a small geographic range, occupies dense forest, and is little known, though observations suggest it has the same territorial habits as the other sunangels. It prefers to feed when perched or clinging, and tends to hold its wings raised for a moment after alighting and when involved in territorial disputes. It rarely strays from thick cover and feeds at low levels, often piercing flowers with long corollas such as *Psammisia* species. The species is thought to be declining.

DISTRIBUTION Southwest Colombia, northwest Ecuador

HABITAT Hilly forest with thick, bushy understory, scrubby forest edges; 3,950–9,200 ft (1,200–2,800 m)

SIZE Length: 3⅞–4⅜ in (10–11 cm). Weight: 5.5 g

STATUS Least Concern

MALE

Tourmaline Sunangel

Another purple-throated green sunangel, with
a rather long and forked tail, the Tourmaline
Sunangel is a darker, richer green than similar
species, with a glittering, dark violet throat
patch, and no white breast band. The female's
throat is whitish with some dark speckling and
sometimes a hint of violet—there is much
variation. In good light the male's forehead
shines a brighter green than the rest of the
plumage. The species is an aggressive, territorial
hummingbird, feeding mainly at low levels.
It usually clings to the flower from which it
is feeding, holding its wings out and raised.
Flowers favored include fuchsias and salvias.
It also spends time picking insects from
vegetation, and perching quietly on the
tops of bushes. It occupies a wide
variety of habitat types, and so is
quite a common species within
its relatively narrow
geographical range.

MALE

FEMALE

DISTRIBUTION The Andes in Colombia and eastern slopes of the Andes in northwest Ecuador

HABITAT Forest interior and edges, clearings, scrubland; 4,900–11,150 ft (1,500–3,400 m)

SIZE Length: 3⅞–4⅜ in (10–11 cm). Weight: 4–4.5 g

STATUS Least Concern

Little Sunangel

Despite its name, this species is about the same size as other sunangels, though it weighs a little less. A dark green bird with a strongly notched tail, it has a black chin and the male has a vivid yellow-orange throat patch (whitish with dark green speckles and just a little yellow-orange in females). When in flight, the white undertail is eye-catching. There are two subspecies: subsp. *micraster*; and subsp. *cutervensis*, males of which have darker, deeper orange throat patches.

The species may be found at the same sites as the Amethyst-throated Sunangel but in general occurs at higher elevations, especially outside the breeding season. It is a typical sunangel in terms of its territorial behavior, and when not feeding may be seen perched at the tops of bushes, surveying the scene. It is generally quite uncommon, and little is known about the details of its behavior.

MALE

DISTRIBUTION Subsp. *micraster* occurs on east-facing Andes slopes in southeast Ecuador and north Peru; subsp. *cutervensis* occurs in northwest Peru

HABITAT Thick, mossy montane forest and forest edges, scrubby pasture; mainly 7,550–11,150 ft (2,300–3,400 m)

SIZE Length: 3⅞–4⅜ in (10–11 cm). Weight: 3.5–4 g

STATUS Least Concern

Purple-throated Sunangel

The Purple-throated Sunangel is a little larger than most of the other sunangels, with a long, deeply forked tail (shorter in females). The plumage is dark, shining green, the male's ornamented with a deep violet throat patch, a patchy band of turquoise-blue on the breast below, and quite a large, vivid blue forehead patch. The apparent color of these highly iridescent areas varies greatly with light direction. The female's throat is whitish with green flecks. The species has a restricted distribution and altitudinal range, but within that area will make use of many habitats. Territorial and aggressive, it shows a particular liking for eucalyptus trees, and in parts of southern Ecuador is apparently dependent on the non-native Tasmanian Blue Gum (*Eucalyptus globulus*). The nest is a stretchy cup made mainly from plant material, in which the female lays two eggs; breeding behavior is not well studied.

DISTRIBUTION Western slopes of the Andes in southern Ecuador and northern Peru

HABITAT Cloud forest, alder woodland, scrubby areas; 7,050–9,850 ft (2,150–3,000 m)

SIZE Length: 4⅜–4¾ in (11–12 cm). Weight: 5–6.5 g

STATUS Least Concern

MALE

Royal Sunangel

The male of this rare sunangel has entirely dark blue plumage, with subsp. *johnsoni* also featuring strong indigo iridescence on the crown, throat, and upper breast. The female is more like a typical sunangel, with glossy green plumage and a white throat band, but has a deep blue-green tail. The male is very territorial, and is often seen defending clusters of *Brachyotum quinquenerve* flowers. The female prefers different flowers, especially those in the family Ericaceae. Birds feed while perched more often than when hovering, and will feed on the ground. They will also use holes already made in flower bases by flowerpiercers to "steal" nectar. The species is rarely observed and has unusually exacting habitat requirements. It is currently known from eight sites, and is declining, as its habitat is rapidly being destroyed—the remaining habitat patches urgently need protection. It has often been seen in areas regularly disturbed by fire.

MALE

DISTRIBUTION Subsp. *regalis* is occurs in Cordillera del Cóndor (northern Peru); subsp. *johnsoni* is found in Cordillera Azul (northern Peru)

HABITAT Elfin scrub, comprising stunted forest and grassland; 4,450–7,200 ft (1,350–2,200 m)

SIZE Length: 4⅜–4¾ in (11–12 cm). Weight: 3.5–4.5 g

STATUS Endangered

Sephanoides sephanoides

FEMALE

Green-backed Firecrown

A fairly small hummingbird, this species is shining olive green on the upperparts with a green-speckled off-white belly, breast, and throat. Only the male has the bright orange crown that gives the firecrown hummingbirds their name. This is a territorial species, though large groups may be found around trees laden with flowers, and will fearlessly chase other, much larger birds. It is a common bird in gardens, parkland, and other non-natural habitats, and is easy to observe, its high metallic *"zip"* call often drawing attention to its presence. The nest is a tiny cup fixed to a branch, sometimes overhanging water. It is the only hummingbird species over much of its range, though it shares Robinson Crusoe Island (of the Juan Fernández Islands) with its endangered relative the Juan Fernandez Firecrown. The most southerly breeders migrate northeast in the austral winter.

DISTRIBUTION The southern tip of South America, from Chile and western Argentina (farther east in winter only) to Tierra del Fuego, also Juan Fernández Islands

HABITAT Forest glades and edges, gardens and other well-vegetated but fairly open habitats; 0–6,550 ft (0–2,000 m)

SIZE Length: 3⅞–4⅛ in (10–10.5 cm). Weight: 5–5.5 g

STATUS Least Concern

Juan Fernandez Firecrown

The male Juan Fernandez Firecrown has a spectacular rich orange plumage and a glittering yellow-orange crown. The female is much smaller and slighter, with green upperparts, dark-spotted pale underparts, and a touch of glossy blue on the crown and rump. There were formerly two subspecies, subsp. *fernandensis* and subsp. *leyboldi*, but the latter, which occurred only on Alejandro Selkirk Island, has been extinct since 1908. This is a territorial feeder, and is a highly adapted specialist pollinator of native plants, especially *Dendroseris litoralis* and *Rhaphithamnus venustus*. Its declining population (about 2,000 individuals) is suffering the effects of habitat loss and the arrival of invasive non-native plants. Surveys also indicate a strong imbalance in the sexes, with three males for every female, and introduced predators such as rats may have an impact. Proactive management of vegetation, by removing invasives and replanting native flora, will be key to saving the species.

FEMALE

MALE

DISTRIBUTION Robinson Crusoe Island (Juan Fernández Islands, off the west coast of Chile)

HABITAT Sheltered, shady woodland, thickets, and gardens; 0–2,950 ft (0–900 m)

SIZE Length: 4⅛–4¾ in (10.5–12 cm). Weight: 7–11 g

STATUS Critically Endangered

Green Thorntail

Like other thorntails, the male of this species has elongated tail feathers, but these are much shorter than in the Wire-crested and Black-bellied thorntails. The plumage is otherwise shining green with a glittering throat patch, and the belly is shining blue-green. There is also a tint of blue on the tail. The short-tailed female has a white cheek stripe and both sexes have white rump bands. Several individuals of these hummingbirds may gather around flowering *Inga* or *Mimosa* trees, and when not taking nectar they pick insects and spiders from the undersides of leaves in the canopy. As is typical for the genus, this species does not adapt to modified habitats, and so is threatened by habitat loss, though its current population trend is not known. It can be found in various protected areas, including Tapantí National Park in Costa Rica.

DISTRIBUTION Costa Rica, Panama, western Colombia, western Ecuador

HABITAT Humid forest, forest edges and clearings; 0–4,600 ft (0–1,400 m)

SIZE Length: 2½–3⅞ in (6.5–10 cm). Weight: 3 g

STATUS Least Concern

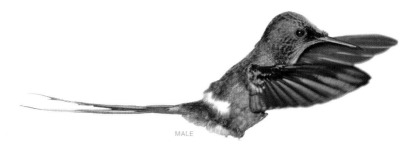

MALE

Black-bellied Thorntail

This species resembles the Wire-crested Thorntail but is uncrested. The male has very elongated, fine outer tail feathers, and a fine chestnut band above the black belly. The bill is also black, and the throat shining green. The female is green with a white lower belly; both sexes have a narrow white rump band. Subsp. *melanosternon* is a little smaller and darker than the nominate. Like other thorntails, this bird feeds mainly in the canopy and has a weaving, insect-like flight. Males in courtship display fan their tails and make a loud cracking sound as they dart to and fro in front of and over the female. The nest is situated high in a tree, on a horizontal branch. Given the high rate of deforestation in its range and its unwillingness to use modified habitats, this species may well be declining.

DISTRIBUTION Subsp. *langsdorffi* occurs in eastern Brazil; subsp. *melanosternon* ranges from southeast Colombia and south Venezuela to eastern Ecuador, east Peru, and west Brazil

HABITAT Humid forest and forest edges; 350–1,000 ft (100–300 m)

SIZE Length: 3–5⅜ in (7.5–13.5 cm). Weight: 3 g

STATUS Least Concern

MALE

Wire-crested Thorntail

The male of this small hummingbird has a
remarkable crest in which a few crown feathers
are greatly extended as fine, wire-like filaments.
The outer tail feathers are also elongated and
narrow, with successive feathers towards the
center of the tail becoming shorter. The male
is otherwise greenish with a glistening throat
patch, black breast and belly, and a narrow
white rump band. The female is similar but
lacks the crest, is short-tailed, and has a bold
white cheek stripe. The species feeds mainly in
the canopy, preferring the flowers of *Inga* trees,
and has a rather bumblebee-like flight. The
nest is also located quite high in the forest
strata, and is a saddle-shaped structure built on
a branch tip. This species' range in the Amazon
Basin is undergoing rapid deforestation, and
as a result the bird is likely to be suffering a
significant population reduction, hence its
listing as Near Threatened.

FEMALE

MALE

DISTRIBUTION Eastern Colombia, east Ecuador, northwest Peru

HABITAT Humid forest and forest edges; 1,650–3,950 ft (500–1,200 m)

SIZE Length: 3–4⅜ in (7.5–11 cm). Weight: 2.5 g

STATUS Near Threatened

Tufted Coquette

The male of this small coquette has a pointed
orange crest and a neck ruff of long, black-
tipped orange plumes. His plumage is olive
green overall, with a glittering green forehead
and throat, a white rump band, and orange tail
sides. The female lacks the crest and plumes,
and has an orange face and breast. A widely
roaming trapliner, this bird flies in the typical
floating, slow, bumblebee-like manner of its
genus and often sneaks into larger species'
territories unnoticed, as well as visiting low-
growing flowers too small to be worthwhile to
larger hummingbirds. It nests in the dry season,
building a small cup nest from fine plant
material. Incubation takes 13 to 14 days, and
fledging a further 20 days. This species is quite
common in parts of its range and may make
seasonal movements, including between the
island of Trinidad and Venezuela.

DISTRIBUTION Trinidad, eastern
Venezuela, the Guianas, north Brazil

HABITAT Forest edges, gallery
forest, somewhat modified habitats;
350–3,300 ft (100–1,000 m)

SIZE Length: 2½–2¾ in (6.5–7 cm).
Weight: 2.5–3 g

STATUS Least Concern

MALE

Frilled Coquette

This is a small hummingbird with green upperparts and gray-green underparts, a spectacular red crest, a bright green forehead, and a sparkling green throat. It has white feathers fanning out from its cheeks, with red patches at their bases, and a short, straight bill with a red base. There is a white band across the bird's rump, and the green tail feathers are relatively short and squared off. The female does not have the red crest and white cheek feathers, but does have orangey and creamy-white patches on the throat. Frilled Coquettes fly from plant to plant feeding on the flower nectar of many different small species, including *Hibiscus* and *Eucalyptus*. They also feed on insects and small spiders. They make a cup-like nest from fine plant fibers, moss, and lichen, which they hang over branches about 10 to 16 feet (3 to 5 m) above ground. The current global population and population trend are unknown, and while the species is regarded as uncommon, it has a large range.

DISTRIBUTION Central and eastern Brazil

HABITAT Lowland areas, humid forests and forest edges, plantations; 0–3,300 ft (0–1,000 m)

SIZE Length: 2¾–3⅛ in (7–8 cm). Weight: 2–3 g

STATUS Least Concern

FEMALE

Rufous-crested Coquette

This is a small hummingbird with green upperparts and underparts. It has a bright crest of long red feathers with dark green tips, a bright green throat and cheek feathers, and pointed white feathers below its collar. The bill is short and straight with a black tip. There is a white band across the bird's rump, and the green tail feathers are rounded, forming a distinctive double tip. The central two tail feathers are green and the outer feathers are orange-red with green edges and black tips. The female does not have the red crest and green cheek feathers, and has a pale reddish throat with duller underparts. The male of subsp. *lessoni* has more pointed crest feathers with hardly any green on their tips, and less pointed throat feathers. Rufous-crested Coquettes feed on the flower nectar of small species such as *Inga*, Verbenaceae, and Myrtaceae, and also take insects and small spiders. They make a cup-like nest from fine plant materials. The current global population and population trend are unknown, and while the species is regarded as rare, it has a large range.

MALE

DISTRIBUTION Subsp. *delattrei* occurs in eastern Peru and northern Bolivia; subsp. *lessoni* occurs in southwestern Costa Rica and Panama, and the central and eastern Andes of Colombia

HABITAT Lowland and mountainous areas, humid forests and forest edges, damp vegetation, semi-open areas, roadside verges; 1,650–7,200 ft (500–2,200 m)

SIZE Length: 2⅜–2¾ in (6–7 cm). Weight: 2–3 g

STATUS Least Concern

Spangled Coquette

The male Spangled Coquette has a long, shaggy orange crest, each feather bearing a large green spot at the tip. The crest may be held fanned out and erect. However, his white-tipped "whiskers" are short and less noticeable than those of most coquettes. Both sexes are otherwise shining green, and the female's face is orange. Reliably separating females of this species from those of the very closely related Rufous-crested Coquette may not be possible. This tiny coquette is a rare and little-known bird, patchily distributed in its narrow range, and has in the past been considered to be Near Threatened because of its apparently declining population. It is most often observed feeding from the blossoms of flowering trees at quite high levels, and will also catch flying insects. It has the steady, rather bumblebee-like flight style typical of the coquettes.

FEMALE

DISTRIBUTION Western Venezuelan coast and in a narrow bank through Colombia and Ecuador to northern Peru

HABITAT Scrubland, forest edges, clearings; 0–4,250 ft (0–1,300 m)

SIZE Length: 2½–2¾ in (6.5–7 cm). Weight: unknown

STATUS Least Concern

Festive Coquette

One of the larger coquettes, this bird looks dark in most lights, apart from its striking whitish rump band. Seen in good light, it shows a shining, dark green head and body, with darker wings. The male has a ruff of short, white-tipped "whiskers" and a dark copper-violet tail; the female has a paler belly and tail. There are three subspecies: subsp. *chalybeus*; the darker subsp. *klagesi*; and subsp. *verreauxii*, in which the male bears a long, pointed crest. The Festive Coquette feeds from many different flower types, and is most often seen visiting flowering trees. It will also pick insects and spiders from plant leaves. It makes a small cup nest on top of a horizontal branch, usually 6 to 16 feet (2 to 5 m) above ground level, and in this the female incubates her two eggs for 13 to 14 days. The chicks fledge at 22 days old.

MALE

DISTRIBUTION Subsp. *chalybeus* occurs in southeast Brazil; subsp. *klagesi* occurs in southeast Venezuela; subsp. *verreauxii* occurs in eastern Colombia, Ecuador, and Peru across to northwest Brazil and central Bolivia

HABITAT Humid forest and secondary forest, cerrado; 350–3,300 ft (100–1,000 m)

SIZE Length: 3–3⅜ in (7.5–8.5 cm). Weight: 2.5–5 g

STATUS Least Concern

White-crested Coquette

With his glittering coppery-orange forehead, white crown that extends to a pointed crest, and very fine, dark "whiskers," the male of this species is quite unlike any other coquette. The female has a white chin and breast; otherwise both are typical coquettes in plumage. This species is primarily a canopy feeder and, like the related Black-crested Coquette, is a trapliner without territorial inclinations and tends to be dominated by other hummingbird species. Resting birds may perch at lower levels and can be very tolerant of close approach by observers. The nest is a cup made of soft plant material such as downy seeds, decorated with lichens and built at a twig tip 15 to 65 feet (5 to 20 m) up in the trees. Chicks fledge at 21 to 22 days old, and both young males and females have a female-like plumage.

DISTRIBUTION Central Costa Rica to west Panama

HABITAT Humid forest, shady plantations, and other well-vegetated, damp habitats; 1,000–3,950 ft (300–1,200 m)

SIZE Length: 2¾–3⅛ in (7–8 cm). Weight: 2.5 g

STATUS Least Concern

MALE

Black-crested Coquette

A smartly marked, small coquette, this bird
has mainly green plumage but a whitish belly
(and throat in the female) marked with shining
bronze speckles, and a whitish-brown rump
band. The male has a long crest of fine, hair-
like dark feathers, and long cream-colored
"whiskers," while the female has a buff tail tip.
This hummingbird feeds by traplining, but also
persistently and discreetly visits the territories
of more dominant species. Its tiny size, plumage,
and stance in flight give it an appearance similar
to an *Aellopos* hawkmoth—territorial birds rarely
chase off insects, so the coquette may benefit
from this resemblance. It often rests on
prominent and exposed bare branches. It feeds
at all levels, but what little is known about its
breeding behavior suggests it prefers to nest
10 feet or more (several meters) up in the trees,
building a small cup nest at a branch tip.

MALE

FEMALE

DISTRIBUTION Southern Mexico,
south to eastern Costa Rica

HABITAT Light forest, forest edges
and clearings, plantations;
350–3,950 ft (100–1,200 m)

SIZE Length: 2½–2¾ in (6.5–7 cm).
Weight: 2.5–3 g

STATUS Least Concern

Ecuadorian Piedtail

Both sexes of this small, mainly bright green hummingbird are alike. It has dark wings and a white belly and throat band, and a white tip and base to its otherwise black wedge-shaped tail, which gives a distinctive banded appearance in flight. It is a very vocal and active bird but can still be difficult to observe well, tending to forage 6 to 13 feet (2 to 4 m) up in the trees. It often clings to flowers when feeding, rather than hovering, and also picks insects from leaves. The well-concealed nest is a tiny cup built among vines, usually well above head height. It is thought the species moves to higher altitudes after breeding. Formerly considered to be Near Threatened, its status was changed to Vulnerable in 2012 because of the high rate of deforestation in its habitat, though at present it is still quite common in some areas.

DISTRIBUTION A narrow band from southern Colombia through east Ecuador to northeast Peru

HABITAT Forest edges, secondary forest; 1,650–3,950 ft (500–1,200 m)

SIZE Length: 2¾–3 in (7–7.5 cm). Weight: 2–2.3 g

STATUS Vulnerable

MALE

Speckled Hummingbird

This hummingbird looks rather similar
to the mountain-gems, having that group's
conspicuous dark cheeks and contrasting white
line behind the eye. The plumage is shining
green above and whitish to buff below with
variable amounts of dark, scaly speckling. The
bill is short and dark, and the legs and feet
reddish. The eight subspecies
vary mainly in their
underside shading and
amount of speckling. This is primarily a solitary
trapline feeder, working its way from flower to
flower in the forest understory, and sometimes
piercing flowers to steal nectar. The nest is
built below a protective overhanging structure,
and is a bulky moss cup lined with soft plant
fibers. The incubation and fledging periods are
17 to 18 days and 20 to 24 days, respectively.
This is a very widespread and, in many areas,
very common montane species, found in several
protected areas.

MALE

DISTRIBUTION Subsp.
melanogenys occurs in the Andes
of west Venezuela and eastern
Colombia through to south-central
Peru; subsp. *debellardiana* is found
on the Colombian–Venezuelan
border; subsp. *aeneosticta* is found
in north and central Venezuela;
subsp. *cervina* occurs in the west
and central Colombian Andes;
subsp. *connectens* is found in Huila
in south Colombia; subsp. *maculata*
ranges from Ecuador to north Peru;
subsp. *inornata* ranges from
southern Peru to northwest
Argentina

HABITAT Humid forest and forest
edges, riverine forest; 3,300–
8,200 ft (1,000–2,500 m)

SIZE Length: 3⅜–3½ in (8.5–9 cm).
Weight: 4–5 g

STATUS Least Concern

Long-tailed Sylph

The male of this beautiful hummingbird has a greatly elongated tail, which may make up well over half the bird's total length. The body plumage is shining green, with a blue throat patch in some forms (and a speckled white throat and orange belly in females). The tail may be green, violet-blue, or a combination of both depending on the subspecies (of which there are six), and the outermost feathers are gently upcurved in most forms. This short-billed species visits short-flowered canopy species like *Inga*, and sometimes steals nectar by flower-piercing. It may trapline, or defend a territory where suitable flowers are clustered. Males have been observed displaying ritualized nest-building behavior, though as with other hummingbirds the nest proper is made by the female alone, which may roost in the domed structure year-round. The species is an altitudinal migrant, and while it is common over much of its range, it is declining.

DISTRIBUTION Subsp. *kingii* is found in the eastern Andes of Colombia; subsp. *margarethae* occurs in north-central and coastal Venezuela; subsp. *caudatus* occurs in west Venezuela and north Colombia; subsp. *emmae* occurs in the central Andes of north Colombia, the western Andes of south Colombia, and northwest Ecuador; subsp. *mocoa* occurs in the central Andes of south Colombia, Ecuador, and north Peru; subsp. *smaragdinus* occurs in the eastern Andes of Peru and west-central Bolivia

HABITAT Forest edges, scrub, light woodland, gardens, plantations; 2,950–9,850 ft (900–3,000 m)

SIZE Length: 3¾–7½ in (9.5–19 cm). Weight: 5–9 g

STATUS Least Concern

MALE

FEMALE

Aglaiocercus coelestis

Violet-tailed Sylph

This spectacular species resembles the Long-tailed Sylph but has a violet-flushed rump. The male's tail feathers are violet and all but the outer (longest) pair have green tips. Subsp. *coelestis* has a shining blue throat patch, while that of subsp. *aethereus* is green. The short, straight bill is thicker towards the tip. Females of both forms have green upperparts, a yellowish belly, a white breast, and a white throat heavily spotted with green. This sylph usually forages near the ground (but also visits *Inga* and *Erythrina* flowers in the canopy) and is principally a trapliner; its distribution shifts through the year as it follows the flowering seasons of its preferred species. Both sexes build nests for roosting, though only the female builds the breeding nest. Incubation and fledging periods are 15 to 17 days and 26 to 30 days, respectively. This is a common bird, with dense populations in a number of protected areas.

MALE

DISTRIBUTION Subsp. *coelestis* occurs on Pacific slopes of the western Andes in Colombia and north and central Ecuador; subsp. *aethereus* is found in southwest Ecuador

HABITAT Cloud forest and more open scrubby country; 3,300–6,550 ft (1,000–2,000 m)

SIZE Length: 3¾–8¼ in (9.5–21 cm). Weight: 4.5–5 g

STATUS Least Concern

Red-tailed Comet

A spectacular hummingbird, this species is named for the male's very long, broad, forked orange-red tail, which makes up nearly half of his total body length. The tail feathers are graduated and each is tipped black, creating a "ladder" of black bands. The red coloration extends to cover the back, becoming deeper farther up; the plumage is otherwise mainly green. Females have much shorter tails and less extensive red. There are two subspecies: subsp. *sparganurus*; and the lighter-colored subsp. *sapho*. The male is territorial, defending a flower-rich patch from other hummingbirds with a dashing, tail-flicking flight. The female builds a rather solid and bulky cup-shaped nest from various plant materials, often sheltered by a rocky overhang, and incubates her two eggs for 19 to 20 days. The chicks fledge at about 31 days. This hummingbird is fairly common and its population is stable.

DISTRIBUTION Subsp. *sparganurus* occurs in north and central Bolivia; subsp. *sapho* occurs in south Bolivia to north and west Argentina and just into eastern Chile

HABITAT Open upland and mountainous countryside with scattered trees, light woodland; 4,900–13,100 ft (1,500–4,000 m)

SIZE Length: 4¾–7⅞ in (12–20 cm). Weight: 5–6 g

STATUS Least Concern

MALE

Bronze-tailed Comet

A medium-sized, rather drab hummingbird with a forked tail, this species has a shining purple-red throat (the female has orange discs instead). The plumage is otherwise iridescent olive green, with a paler belly and a bronze gloss to the longer tail feathers. It is the only species in the genus *Polyonymus* but is thought to be quite closely related to the Gray-bellied Comet, which is itself the only species in its genus (*Taphrolesbia*). It feeds on various plants, including *Agave* and several cactus species, and is regularly driven away from good food sources by other more aggressive species. It is a little-known species and has previously been classified as Near Threatened, but was downgraded to Least Concern in 2004 as its population has been quantified at more than 10,000 individuals and its numbers now seem to be stable.

DISTRIBUTION South and central Peru on the Pacific slope of the western Andes

HABITAT Scrubland and woodland edges on mountain slopes; 4,900–11,800 ft (1,500–3,600 m)

SIZE Length: 4⅜–5⅛ in (11–13 cm). Weight: 5–5.5 g

STATUS Least Concern

FEMALE

Gray-bellied Comet

A dark, elegant, medium-sized hummingbird with a deeply forked tail, this species has shining, dark green upperparts and gray underparts, and the male has an iridescent blue throat patch. The bill is rather short. This is a very little-known species, which has only been observed a handful of times since the 1950s. Its population is estimated to be no more than 1,000 individuals, and because it seems to have disappeared from some previously occupied sites it is thought to be declining, hence its designation of Endangered. It is described as territorial and aggressive, dominating all other hummingbirds that share its habitat. The nest is built under a rocky overhang for shelter, and may be used repeatedly for successive broods. Habitat destruction (burning scrubland to create pasture) is ongoing within its known range—thorough population surveys and habitat protection are urgently needed.

DISTRIBUTION Known from nine small, scattered sites in the northwestern Andes of Peru

HABITAT Scrubby, dry, rocky, mountainous country; 8,550–11,500 ft (2,600–3,500 m)

SIZE Length: 5½–6¾ in (14–17 cm). Weight: unknown

STATUS Endangered

MALE

Ecuadorian Hillstar

A medium-sized hummingbird, the Ecuadorian Hillstar has a rather short, slightly downcurved bill and, in the male, a shining violet-blue head and green lower throat. The female is green-headed, and both sexes have a white belly, green back, dark wings, and a dark tail with white flashes. There are three subspecies: subsp. *chimborazo*; subsp. *jamesonii*, in which the male's throat is entirely violet; and subsp. *soderstromi*, in which the green throat patch is much reduced. An aggressive and territorial bird, it clings on flowers (especially *Chuquiraga insignis*) when feeding, and in bad weather conserves energy by resting quietly in sheltered spots. At night it roosts in caves and becomes torpid to survive the low temperatures. The well-hidden nest is large, bulky, and thickly lined with soft materials for heat conservation during the long incubation and chick-rearing periods (about 20 and 38 days, respectively).

MALE

DISTRIBUTION Subsp. *chimborazo* occurs on Mt. Chimborazo, central Ecuador; subsp. *jamesonii* occurs in mountains of the far south of Colombia and north Ecuador; subsp. *soderstromi* occurs on Mt. Quilotoa, central Ecuador

HABITAT Puna grassland and scrubby paramo on high mountainsides; 11,500–17,050 ft (3,500–5,200 m)

SIZE Length: 5⅛ in (13 cm). Weight: 8 g

STATUS Least Concern

FEMALE WITH CHICKS

Andean Hillstar

The male of this hillstar is dull olive-brown above and white below, with a brown belly stripe and a shining green throat. Females have mottled white throats and gray-green undersides. There are three subspecies: subsp. *estella*; subsp. *stolzmanni*, which has a black rather than brown ventral stripe; and subsp. *bolivianus*, which has black spots on its belly stripe. Adapted to living at high-altitudes, the Andean Hillstar feeds primarily from *Chuquiraga spinosa* and perches rather than hovers when taking nectar. It also becomes torpid overnight. The large, thickly lined cup nest is stuck with saliva to rock surfaces, usually angled to catch warmth from the rising sun; the best spots in sheltered rocky gorges attract several females to build in close proximity. The species has an exceptionally high nesting success rate and may rear two broods before moving downslope for the austral winter. It is one of the most common birds in its preferred puna habitat.

DISTRIBUTION Subsp. *estella* occurs in the Andes of southwestern Peru, west Bolivia, northern Chile, and northwest Argentina; subsp. *stolzmanni* is found from southern ecuador to central Peru; subsp. *bolivianus* occurs in the Andes of central Bolivia

HABITAT Vegetated mountainsides (especially puna grassland), woodland edges, around towns and villages; 7,850–16,400 ft (2,400–5,000 m)

SIZE Length: 5⅛–5⅞ in (13–15 cm). Weight: 8–9 g

STATUS Least Concern

MALE

FEMALE

Oreotrochilus leucopleurus

White-sided Hillstar

The White-sided Hillstar is the southernmost species of hillstar. A rather drab species, particularly the female, it is very similar to the closely related Andean Hillstar but the male has a very broad black belly patch. Its distribution overlaps with that of Andean Hillstar, and where the two coexist they sometimes hybridize. The White-sided shows the typical hillstar adaptations to living at high altitude, including entering a torpid state overnight, with greatly reduced heart rate and body temperature. It feeds preferentially from *Chuquiraga* bushes, perching rather than hovering to save energy. The large nest is anchored to a vertical rock face in a well-sheltered location, and incorporates soft material such as feathers and animal hair for insulation. The species moves down to lower altitudes after the breeding season. Though not very well studied, it is generally a common hummingbird, and its population is currently stable despite some ongoing habitat loss over much of its range.

DISTRIBUTION The Andes through southern Bolivia to south-central Chile and south-central Argentina

HABITAT Scrubby puna grassland; 3,950–13,100 ft (1,200–4,000 m)

SIZE Length: 5⅛–5⅞ in (13–15 cm). Weight: 8–8.5 g

STATUS Least Concern

MALE

Black-breasted Hillstar

The female Black-breasted Hillstar looks fairly typical for the genus, though with reduced white in the tail. The male is very distinctive, with a velvety black breast and belly, and contrasting light grayish-orange flanks. There has been speculation that this is a melanistic variant of the Andean Hillstar rather than a full species. Like other hillstars it prefers to feed from the flowers of *Chuquiraga* bushes, in this case *C. spinosa*, but also visits flowering cacti and searches among the grasses for smaller, mainly red-petaled flowers, and hawks for insects high in the air. It fixes its bulky nest on a sheltered vertical rock face, or sometimes on the wall of a building. Within its rather small range it is a common bird and occurs in several protected areas—in general, its habitat is not under significant threat of destruction.

DISTRIBUTION A few mountainsides in central Peru

HABITAT Puna grassland on mountain slopes, rocky scrubland, gardens; 11,500–14,450 ft (3,500–4,400 m)

SIZE Length: 5⅛–5½ in (13–14 cm). Weight: 8.5 g

STATUS Least Concern

MALE

DISTRIBUTION The Andes, from Colombia through eastern Ecuador to north Peru

HABITAT Forest edges and scrubland on mountain slopes; 8,550–11,800 ft (2,600–3,600 m)

SIZE Length: 3½–4 in (9–10 cm). Weight: 6–7 g

STATUS Least Concern

MALE

Opisthoprora euryptera

Mountain Avocetbill

This small hummingbird's short bill is slightly upturned at the tip, though not nearly so much as those of the wading birds for which it is named. It is dark green with a blackish head and wings, and a paler belly marked with dark spots and flushed orange. The sexes are alike in plumage, though males are larger and heavier. It takes nectar from various small tubular flowers, including Ericaceae and *Fuchsia* species, either hovering or clinging while it feeds, and is generally seen fairly low down (3 to 10 feet/1 to 3 m) and alone. It spends much of its time resting and is generally less active than most other hummingbirds that share its habitat. Although not currently considered threatened, it is rather uncommon and is likely to face problems in the future as its habitat is generally unprotected and suitable for agricultural development.

Black-tailed Trainbearer

The male of this rather short-billed, glossy green hummingbird has very long and slightly upcurved tail streamers that make up considerably more than half of his total length. Males also have a glistening green throat patch, while females are considerably shorter-tailed and have green-spotted white underparts. Subsp. *victoriae* is particularly long-tailed; subsp. *juliae* has a shorter tail; and subsp. *berlepschi* has a more downcurved bill than other subspecies. This bird feeds at mainly high levels in the forest strata and will visit introduced as well as native species. It also hawks for insects. The male's long tail makes a snapping sound as the bird performs its courtship flight. The nest is quite bulky and placed below an overhanging shelter of some kind; incubation and fledging takes 18 to 19 days and 29 to 31 days, respectively. This is a common bird with a stable population.

DISTRIBUTION Subsp. *victoriae* occurs in the Andes from northeast Colombia through south Colombia to Ecuador; subsp. *juliae* is found in the Andes of north and central Peru; subsp. *berlepschi* occurs in the Andes of southeast Peru

HABITAT Mountainous forest edges, woodland, gardens, scrub; 8,550–13,100 ft (2,600–4,000 m)

SIZE Length: 5⅞–10¼ in (15–26 cm). Weight: 5 g

STATUS Least Concern

MALE

MALE

Green-tailed Trainbearer

This species is smaller and shorter-tailed than the very closely related Black-tailed Trainbearer. The male's tail makes up about half his total length, and all but the longest tail feathers are mostly green rather than black. Subsp. *nuna* has broader and blacker tail streamers than subsp. *gouldii*, while subsp. *pallidiventris* has very narrow streamers, and subsp. *gracilis* is smaller and more buff-tinted. The species prefers drier habitats than its congener, but there is considerable overlap and some hybridization is known to have occurred—subsp. *huallagae*, which has variously been described as a subspecies of both trainbearers, may in fact be a hybrid between them. The species forages at lower levels than the Black-tailed, and often visits flowers to take pollinating insects from them as well as nectar. It is fairly common and, despite some deforestation in its range, is thought to have a stable population.

DISTRIBUTION Subsp. *nuna* occurs in southern Peru; subsp. *gouldii* is found in northeast and south-central Colombia; subsp. *gracilis* occurs northern and central Ecuador; subsp. *aureliae* is found in southern Ecuador; subsp. *pallidiventris* is found in northern Peru; subsp. *huallagae* occurs in central Peru; subsp. *boliviana* is found in northwest and central Bolivia

HABITAT Scrub, paramo, secondary growth; 5,600–12,450 ft (1,700–3,800 m)

SIZE Length: 4½–6¾ in (11.5–17 cm). Weight: 4 g

STATUS Least Concern

FEMALE

Black-backed Thornbill

DISTRIBUTION Northeastern Colombia

HABITAT Mountainous areas, forests, forest edges, paramo, bushy areas at higher elevations by the treeline; 6,550–15,100 ft (2,000–4,600 m)

SIZE Length: 3½–3⅞ in (9–10 cm). Weight: 3–4 g

STATUS Endangered

The Black-backed Thornbill has a dark, velvet black head and back, and mottled green, gray, and orange breast and underparts. It has small white patches behind each eye, a throat patch of bright, shining green feathers, and a short black bill that is slightly downcurved. Its long, forked tail feathers are purple-black, and the uppertail coverts are purple-bronze. The female has green upperparts and white underparts with green mottling, and her tail is slightly shorter, with rounded tips. The bird feeds on the nectar of a range of species at all heights, often clinging to the flowers. It also takes insects and small spiders, sometimes catching them in flight. Little is known about its breeding behavior. Birds move to lower elevations in May and June. The current global population is unknown, although it does appear to be in decline as a result of severe habitat loss and fragmentation, and the species is regarded as uncommon in its small range.

MALE

Purple-backed Thornbill

The Purple-backed Thornbill has a dark, shining purple head and back, and green underparts. It has small white patches behind each eye, a throat patch of bright, shining green, and a very short, straight black bill. Its forked tail feathers are dark purple, and it has reddish-bronze undertail coverts with brownish edges. The female has green upperparts, white underparts with green mottling, and a less forked tail. Subsp. *albiventre* has paler edges to its undertail coverts and subsp. *bolivianum* is darker green underneath and its undertail coverts are pale gray and black. The bird is a trapliner, often clinging to flowers as it takes nectar from a range of species growing from above head height up to the canopy. It also feeds on insects and small spiders. The cup-like nest is built on a level branch from fine plant material. The current global population is unknown but appears to be in decline, and the species is regarded as uncommon in its range.

DISTRIBUTION Subsp. *microrhynchum* ranges from western Venezuela to northwestern Peru; subsp. *albiventre* occurs in central Peru; subsp. *bolivianum* is found in Bolivia

HABITAT Mountainous areas, forest edges, semi-open areas, paramo; 4,900–11,500 ft (1,500–3,500 m)

SIZE Length: 3⅛–3⅞ in (8–10 cm). Weight: 3–4 g

STATUS Least Concern

MALE

Rufous-capped Thornbill

DISTRIBUTION Southeast Ecuador along the Andes through Peru to west/central Bolivia

HABITAT Humid forest glades and edges, secondary forest; 4,600–11,500 ft (1,400–3,500 m)

SIZE Length: 4⅛ in (10.5 cm). Weight: 3.5–4 g

STATUS Least Concern

Smaller than other thornbills, this is a compact, short-billed, broad-tailed hummingbird. Both sexes are dark green, with darker wings and tail and a diffuse orange stripe down the breast and belly, but the male's colors are generally brighter and he also has a shining green throat and a chestnut crown. Although many flowers are too long for the species to feed from them directly, it will readily "steal" nectar by puncturing the flower base with its needle-like bill, or by using incisions made by other flower-piercing birds. The male defends a feeding territory, chasing off other hummingbirds, and when feeding both sexes tend to cling rather than hover. The Rufous-capped Thornbill is not an especially common bird and its behavior is not well understood. A record from Pan de Azucár in Colombia may be an error but could indicate that the species is at least partially migratory.

MALE

Chalcostigma stanleyi

Blue-mantled Thornbill

This very dark thornbill has extensive deep violet-blue iridescence on its back and tail, and the male has a green and pink throat patch (reduced to just a small green chin spot on the female). There are three subspecies: subsp. *stanleyi*; subsp. *versigulare*, which has a narrower throat patch and more violet-blue coloration; and subsp. *vulcani*, which has a duller throat patch. The species is a low-level feeder, visiting small flowers and also picking at the sugary secretions of aphids on leaf undersides. It hops on the ground in search of insects. Males are territorial. The species is not thought to be migratory in the strict sense, but it does undertake altitudinal movements in response to changing weather conditions. It is a common bird in portions of its range but is declining, partly because substantial areas of its preferred *Polylepis* woodland are being cleared for agriculture.

MALE

DISTRIBUTION Subsp. *stanleyi* occurs in the Andes in Ecuador; subsp. *versigulare* occurs in the Andes in eastern Peru as far as the Carpish Mountains; subsp. *vulcani* occurs in eastern Peru south of the Huallaga River into west-central Bolivia

HABITAT Scrubland and light woodland on steep mountainsides; 7,200–13,800 ft (2,200–4,200 m)

SIZE Length: 4¾–5⅛ in (12–13 cm). Weight: 4.5–6 g

STATUS Least Concern

Bronze-tailed Thornbill

A fairly large, dark olive-green thornbill, this bird has a narrow, shining green throat patch (edged with rose-pink in the male) and a bronze sheen to the face sides, rump, and broad tail. Young birds have more chestnut tints on the crown. Like other thornbills, this is a low-level feeder, but it has not been observed actually feeding on the ground. However, it gleans insects from foliage and is also a skilled flycatcher, often killing its prey with a snapping bite and then tossing it up into the air to swallow it. When taking nectar, it clings to flowers, and in flight it beats its wings relatively slowly for a hummingbird. It is usually seen alone and is often territorial. Habitat loss and degradation in its home in the northern Andes has caused this already uncommon species to decline further, although it is not yet considered to be at significant risk.

MALE

DISTRIBUTION Far west of Venezuela and eastern Andes in Colombia

HABITAT Paramo scrub and woodland edges on mountainsides; 9,850–12,800 ft (3,000–3,900 m)

SIZE Length: 5⅛–5½ in (13–14 cm). Weight: 6–6.5 g

STATUS Least Concern

Rainbow-bearded Thornbill

This is a small, mostly dark olive-green thornbill, named for the male's spectacular iridescent throat patch, which grades from blue-green through yellow to red. As with other thornbills, the feathers that form this patch are large and stand slightly away from those surrounding them, giving the impression of a beard. Both sexes have a reddish crown that rises to a small crest, and broad white tail tips. There are two subspecies: subsp. *herrani*; and the darker subsp. *tolimae*. The species is an avid insect-hunter, gleaning foliage at low levels and also flycatching on the wing. The nectar sources it uses most are *Puya* and *Brachyotum* flowers, to which it clings, wings beating, as it feeds. It is thought to be declining. Its range includes several key protected areas, though one of these, the Podocarpus National Park, is under pressure from would-be gold prospectors.

DISTRIBUTION Subsp. *herrani* occurs in the western Andes of Colombia; subsp. *tolimae* occurs on Nevado de Tolima in central Colombia

HABITAT Scrubland and woodland on mountain slopes; 8,850–13,100 ft (2,700–4,000 m)

SIZE Length: 3⅞–4¾ in (10–12 cm). Weight: 5.5–6.5 g

STATUS Least Concern

MALE

Bearded Helmetcrest

A rather small, mainly green hummingbird with a short bill and shortish tail, this species is striking because of its bold white facial markings and, in the male, a long, spiky black-and-white crest and a shaggy, green-centered white "beard." There are four subspecies: subspp. *guerinii, cyanolaemus, lindenii,* and *stubelii.* The first three vary slightly in size and beard color, while subsp. *stubelii* has creamy-yellow rather than white facial markings and a much reduced "beard." It is rather thornbill-like in its behavior, taking many insects in flight and also on the ground, and clinging rather than hovering when taking nectar. The bulky nest is built near water. Incubation takes 21 to 23 days and the chicks grow slowly, as is common at high altitudes, fledging at 35 to 38 days old. This species is common in some parts of its range but is declining.

DISTRIBUTION Subsp. *guerinii* occurs in the eastern Andes of Colombia; subsp. *cyanolaemus* occurs in northeast Colombia; subsp. *lindenii* occurs in the Andes of northwest Venezuela; subsp. *stubelii* occurs in central Colombia

HABITAT Open paramo scrub with rocky gorges, woodland edges; 9,850–17,050 ft (3,000–5,200 m)

SIZE Length: 4⅜–5⅛ in (11–13 cm). Weight: 4.5–6 g

STATUS Least Concern

MALE

Bearded Mountaineer

A rather large hummingbird with a long bill and long, deeply forked, white-edged tail, the male of this species is mainly brown above and white below but has a short violet crest and long, violet-tipped green "beard." The shorter-tailed female has a much smaller, all-green "beard" and her body plumage is more green than brown. There are two subspecies, subsp. *nobilis* and subsp. *albolimbata*, the latter with extensive white scaling on a dark chestnut crown. Like many other high-altitude hummingbirds, this poorly known and rather uncommon species tends to cling to flowers when feeding, but also sometimes hovers below them, constantly spreading and closing its tail. It is often seen visiting flowers of the introduced *Nicotiana* (tobacco plants), and catches many insects. Other hummingbirds dominate it, including some smaller species. It can survive in quite degraded habitat and its population is stable.

DISTRIBUTION Subsp. *nobilis* occurs in the Andes of south-central Peru; subsp. *albolimbata* occurs in central-western Peru

HABITAT Scrubland and light woodland on mountainsides, often near rocky gorges; 8,200–12,450 ft (2,500–3,800 m)

SIZE Length: 5½–6¾ in (14–17 cm). Weight: 9 g

STATUS Least Concern

MALE

Tyrian Metaltail

This typical metaltail has a short, needle-like bill. The male has a glittering green throat and is otherwise rather olive green apart from his tail, which varies between subspecies: It is reddish in subsp. *tyrianthina*, violet in subsp. *districta*, copper red in subsp. *chloropogon*, dark red in subsp. *oreopola*, bronze-green in subsp. *quitensis*, purple-blue in subsp. *septentrionalis*, and bright blue in subsp. *smaragdinicollis*. The female has whitish underparts with a green-scaled belly and a warm buff flush to the throat and upper breast. The male Tyrian Metaltail is aggressive and territorial, though it is not uncommon to see groups feeding in fairly close proximity around flowers like *Eugenia*, *Palicourea*, and *Salvia*. Birds will pierce longer flowers to steal nectar, and hawk for insects. The nest is well hidden among moss or roots, and may have a partial roof for shelter. This bird is widespread and common, with a stable population.

FEMALE

MALE

DISTRIBUTION Subsp. *tyrianthina* is found across the Andes from northwest Venezuela through Colombia and east and central Ecuador to north Peru; subsp. *districta* occurs on the Sierra Nevada de Santa Marta and Serranía du Perijá in Colombia; subsp. *chloropogon* is found on Caribbean slopes in north Venezuela; subsp. *oreopola* occurs in the Venezuelan Andes; subsp. *quitensis* is found in northwest Ecuador; subsp.

septentrionalis occurs in the Peruvian Andes; subsp. *smaragdinicollis* is found in the eastern Andes of Peru and north Bolivia

HABITAT Open damp forest and forest edges, elfin forest, occasionally scrub; 4,900–13,800 ft (1,500–4,200 m)

SIZE Length: 3½–3⅞ in (9–10 cm). Weight: 3.5 g

STATUS Least Concern

Viridian Metaltail

This vivid green metaltail shows considerable subspecific variation across its quite limited distribution. The four subspecies are *williami*, *recisa*, *primolina*, and *atrigularis*, which are all sometimes considered full, separate species. The nominate subsp. *williami* has a blue rather than green tail; subsp. *recisa* has a shorter bill and more forked tail; subsp. *primolina* has a reddish-black upperside to its tail; and subsp. *atrigularis* has a black throat patch. Females of all forms have paler undersides with dark speckling. The Viridian Metaltail is most often observed hover-feeding at the flowers of Ericaceae or Melastomataceae shrubs, and males are territorial. Females may nest in loose colonies, building their moss-based nests in sheltered hollows among rocks or roots. The species is rather uncommon and declining throughout most of its range, as large areas of its habitat have been lost through conversion of scrubland to pasture.

DISTRIBUTION Subsp. *williami* occurs in the central Andes of Colombia; subsp. *recisa* occurs in the Páramo de Frontino in north-central Colombia; subsp. *primolina* occurs in the eastern Andes in south Colombia and north Ecuador; subsp. *atrigularis* occurs in south Ecuador

HABITAT Scrubby paramo grassland, stunted forest; 6,900–13,100 ft (2,100–4,000 m)

SIZE Length: 4⅜–4¾ in (11–12 cm). Weight: 4–5 g

STATUS Least Concern

MALE

Violet-throated Metaltail

This bird's tail is shining blue on the upperside and green below. The plumage is otherwise deep green, with an iridescent purple throat patch in the male. The female's underside is lighter green than the male's and is marked with darker spotting. Within its very specialized habitat the species feeds from a wide variety of plants, primarily at low to mid-level in the strata but sometimes higher. Males are territorial around their feeding grounds. There have been few observations of breeding behavior, but one nest found was a pouch shape and built from moss, twigs, and wool. This hummingbird has a very small population—no more than 2,500 individuals—and is declining. The key threat it faces is habitat loss, as *Polylepis* forest and paramo are burned prior to conversion for agricultural uses. Local support and education are needed to conserve the species.

FEMALE

DISTRIBUTION Cajas Plateau, south-central Ecuador

HABITAT *Polylepis* woodland with lush vegetation, adjoining paramo; 10,350–12,150 ft (3,150–3,700 m)

SIZE Length: 3⅞–4⅜ in (10–11 cm). Weight: 4–4.5 g

STATUS Endangered

Neblina Metaltail

Both sexes of this metaltail are green above and paler with speckles below; the male also has a deep red throat patch, while the female's throat is only lightly speckled with red. The tail is dark blue on the upperside. A hardy species, this hummingbird continues to feed actively in weather conditions bad enough to drive other local hummingbirds to more sheltered spots. The male is territorial, defending clusters of favorite flowers such as *Castilleja*. The female builds a cup nest, using moss as the main material, in a location where there is some shelter from the prevailing wind. Like most metaltails, this species is confined to a small geographic range. It was formerly classified as Near Threatened, but surveys of its habitat in the 1990s revealed that it was quite common in several areas, although it is declining.

DISTRIBUTION South Ecuador and the far north of Peru

HABITAT Scrubland, paramo, and damp, stunted forest; 8,550–11,000 ft (2,600–3,350 m)

SIZE Length: 3⅞–4⅜ in (10–11 cm). Weight: 5–5.5 g

STATUS Least Concern

MALE

Coppery Metaltail

This hummingbird has a shining reddish-green head, upperparts, and underparts, with a slim throat patch of bright, shining green, and a medium-length straight black bill. Its tail feathers are purple on top and gray-purple underneath. The female is similar but has a duller throat patch and white tips on the underside of the outer tail feathers. Subsp. *parkeri* lacks the reddish tones and has a green crown, and its tail feathers are pale blue on top and greenish yellow underneath. The Coppery Metaltail forages at all heights in the forest, and hovers as it takes nectar from the flowers of its preferred plant species, including Melastomataceae and Ericaceae. Little is known about the bird's breeding or behavior. The current global population is unknown but it appears to be in decline as a result of habitat loss, although the species is regarded as fairly common in its range.

DISTRIBUTION Subsp. *theresiae* is found in northern and central Peru; subsp. *parkeri* occurs in northern Peru

HABITAT Mountainous areas, forest edges, clearings, damp grassland, scrubland; 9,850–13,100 ft (3,000–4,000 m)

SIZE Length: 3⅞–4⅜ in (10–11 cm). Weight: 4–5 g

STATUS Least Concern

MALE

141

Scaled Metaltail

A little larger than the other green metaltails, this species has an entirely green tail. The male has a brilliantly iridescent green throat patch, while the female's throat is buff with fine green mottling. There are two subspecies: subspp. *aeneocauda* and *malagae*. The tail of the latter is red on the underside and bronze-toned on the upperside, rather than green. This species feeds from a wide range of flowers with corollas ¾ to 1½ in (2 to 4 cm) long, and often clings to feed. It also takes insects and spiders. The male is territorial around good feeding sites. The bird is most common at the higher end of its altitudinal range, but there is some evidence of dispersal both up and down the slopes after the breeding season. Its adaptability safeguards it to some extent from future habitat modification, but its numbers are thought to be declining.

DISTRIBUTION Subsp. *aeneocauda* occurs on the eastern slopes of the Andes from southeast Peru to northwest Bolivia; subsp. *malagae* occurs in the Incachaca area of central Bolivia

HABITAT Scrubland and glades in cloud forest; 8,200–11,800 ft (2,500–3,600 m)

SIZE Length: 4¾–5⅛ in (12–13 cm). Weight: 5–5.5 g

STATUS Least Concern

MALE

Black Metaltail

The largest of the metaltails, this species has a
shining black head, upperparts, and underparts,
with a hint of red. It has a small, shining
greenish-purple throat patch, small white spots
behind each eye, and a medium-length straight
black bill. In addition, it has white feathers at the
tops of its legs, and a forked bronze-toned tail.
The female is similar but has a smaller throat
patch. The species can mostly be found
hovering at low levels in the understory
and near the ground, where it takes nectar from
a range of plants and trees; it also sometimes
feeds at greater heights in trees. It builds a nest
from moss and roots, suspended from a branch
over a stream. The current global population is
unknown but it appears to be in decline as a
result of habitat loss, although the species is
regarded as fairly common in its range.

DISTRIBUTION Central and
northern Peru; possibly also
into northern Chile and Bolivia,
but this is unconfirmed

HABITAT Semi-open areas
with semiarid woods and
scrubland, mountainous areas
with bushy and wooded
slopes; 4,600–14,750 ft
(1,400–4,500 m)

SIZE Length: 4¾–5½ in
(12–14 cm). Weight: 5.5–6 g

STATUS Least Concern

FEMALE

143

Greenish Puffleg

Formerly considered to be a subspecies of the Buff-thighed Puffleg, this is a small, quite long-billed hummingbird with light yellow-green plumage, darker wings, and a bluish tail. The sexes are similar. The white feathers at the bases of the legs are long, creating "leg puffs" that partly cover the feet when the bird is resting. There are six subspecies, including subsp. *caucensis*, which has a whiter lower belly and more coppery upperparts than the nominate; and subsp. *russata*, which is longer-billed and even more copper-toned. The species forages in the understory of the forest, selecting flowers with short corollas and vigorously defending good clumps of blossoms from other nectar-feeding birds. It will also visit the canopy to feed from *Inga* flowers when they are available. The cup-shaped nest is built beneath an overhanging leaf for protection from rain. Over much of its range this is a common species.

MALE

DISTRIBUTION Subsp. *aureliae* occurs in the eastern Andes of Colombia; subsp. *caucensis* occurs in southeast Panama, south to the west and central Andes of Colombia; subsp. *russata* occurs on the eastern Andean slopes in Ecuador

HABITAT Wet forest on mountainsides; 4,900–10,150 ft (1,500–3,100 m)

SIZE Length: 3½–3⅞ in (9–10 cm). Weight: 4.7–6.5 g

STATUS Least Concern

Hoary Puffleg

This is a rather dark and drab puffleg. Most of the plumage is iridescent olive green, shading to copper on the rump. The throat and breast are sooty gray, darker in the male. The wings and tail are dark, and the small leg puffs are white. This species feeds mainly in the thick understory, where it holds territory and visits flowers with shorter corollas, such as *Palicourea* and *Besleria*. It may also visit flowering trees in the canopy, and most of the insects it takes are gleaned from leaves rather than caught in flight. The nest is ball-shaped, and built on the underside of a hanging leaf to protect it from rain. The species' status of Near Threatened reflects its very small range and declining population, though at present it is still fairly common and easy to find within its range. It occurs in several well-protected reserves.

FEMALE

DISTRIBUTION Southwest Colombia, Pacific slope of northwest Ecuador

HABITAT Cloud forest, forest edges, damp scrub, secondary growth; 3,950–8,200 ft (1,200–2,500 m)

SIZE Length: 3½–3⅞ in (9–10 cm). Weight: 5–6 g

STATUS Near Threatened

Black-breasted Puffleg

This is the darkest and most distinctive of the *Eriocnemis* pufflegs, and the one with the most marked sexual dimorphism. The male is glossy blackish green with a violet vent and throat patch. The female is shining green with dark wings and a green-spotted pale belly, and shares the male's violet markings. Both have white leg puffs. This territorial species mainly feeds from *Palicourea huigrensis* trees but also visits Ericaceae species. It may perch to feed rather than hover. The tiny remaining population (no more than 270 individuals) is under continued pressure due to clearance (for firewood and to make space for ranching) of the forest on which it relies. It may also be threatened by climate change, and by competition with the Gorgeted Sunangel, which is expanding its range. It occurs in protected areas, but careful habitat management is required to ensure its exacting needs are met.

MALE

FEMALE

DISTRIBUTION Northwest Ecuador (mainly on Pichincha and Atacazo volcanoes)

HABITAT Dense elfin forest on mountain ridge crests; 7,850–15,100 ft (2,400–4,600 m)

SIZE Length: 3⅛–3½ in (8–9 cm). Weight: 4.3–4.6 g

STATUS Critically Endangered

Glowing Puffleg

This fairly small puffleg has a short but
deeply forked tail. It is mostly
bright green but with an
iridescent violet vent and chin
patch, and a blackish breast. The leg
puffs are white. The female's chin patch is
much reduced, and she has extensive golden
speckling on her breast and chin. There are four
subspecies: subsp. *vestita*; subsp. *arcosae*; subsp.
paramillo, in which the violet of the chin extends
patchily to the breast; and subsp. *smaragdinipectus*,
which has a lighter green breast. This is a lively
territorial bird that defends clumps of flowers
such as *Palicourea*, and other flowers with short
corollas. It usually feeds at low heights and is
very quick and hence difficult to track. It may
hover or cling when feeding, and catches insects
while in flight. Although it is rather localized
and not especially common, its population is
believed to be stable.

MALE

DISTRIBUTION Subsp. *vestita*
occurs in northwest Venezuela
and the eastern Andes in Colombia;
subsp. *arcosae* ranges from south-
central Ecuador to northwest Peru;
subsp. *paramillo* occurs in the west
and central Andes in Colombia;
subsp. *smaragdinipectus* occurs in
southwest Colombia and Ecuador

HABITAT Open forest
edges, scrubland, grassland;
7,400–12,600 ft (2,250–3,850 m)

SIZE Length: 3½–3⅞ in (9–10 cm).
Weight: 4.4–5.2 g

STATUS Least Concern

Black-thighed Puffleg

This is the only *Eriocnemis* puffleg without white leg puffs—instead, they are black in the male and gray in the female. The bird is otherwise shining green with a dark tail and wings, and with breast speckles that are blackish in the male and white in the female. As with the other *Eriocnemis* pufflegs it is primarily territorial, feeding at low levels from such plants as fuchsias and heathers. It is not a well-studied species but it is considered to be Near Threatened because of its restricted and fragmented distribution, and because of a declining population resulting from habitat loss. It occurs in some protected areas, and locating and protecting other populations is a priority for conservation plans. The species' readiness to use habitats modified by human activity, such as gardens, may mean it will be possible to establish corridors of usable habitat to link up isolated populations.

DISTRIBUTION Central Andes of Colombia and northwest Ecuador

HABITAT Humid scrubland, pasture, gardens, forest edges; 8,200–11,800 ft (2,500–3,600 m)

SIZE Length: 3⅞ in (10 cm). Weight: unknown

STATUS Near Threatened

MALE

Coppery-bellied Puffleg

A shining green puffleg with no chin patch, this species has a violet vent and red-brown belly patch (yellowish in the female). Like others of its genus, this bird is a low-level feeder and is usually territorial around clumps of flowers with short corollas such as *Palicourea* and *Gaultheria*. It darts in quickly to feed, sometimes clinging to the plant as it does so, or to see off an intruder. When resting, it perches in cover, staying alert for any opportunities to hawk for flies. The nest, a fairly substantial cup, is usually hidden in dense vegetation. This hummingbird's conservation status was formerly considered to be Vulnerable but recent surveys have found it to be less scarce than previously thought. However, it remains at risk because its population is declining through ongoing deforestation that is removing or degrading its habitat.

MALE

DISTRIBUTION Northwest Venezuela and the eastern Andes of Colombia

HABITAT Mainly open scrubby and humid habitats, forest edges, sometimes within forest; 6,550–9,850 ft (2,000–3,000 m)

SIZE Length: 3½–3⅞ in (9–10 cm). Weight: 5.6 g

STATUS Near Threatened

Sapphire-vented Puffleg

This is one of the larger pufflegs, and is a gracefully proportioned bird with a long, deeply forked tail. The plumage is very similar in both sexes, being bright, shining green with a darker tail and wings, white leg puffs, a violet-blue vent, and a sky-blue forehead patch. Females have slightly shorter tails and wings. The subspecies differ slightly in the colors of their forehead, crown, and vent. This is a typically aggressive and territorial puffleg. It feeds low down, hovering, clinging, and sometimes even settling on the ground when accessing very low flowers. The nest is well hidden, for both shelter and protection, and is a deep cup made of moss, lichen, and leaf fragments, anchored to a twig on one side. After the two young have been reared, some altitudinal dispersal is likely to take place. In parts of its range this is a very common bird, with healthy populations in several protected areas.

DISTRIBUTION Subsp. *luciani* ranges from central Ecuador to southwest Colombia; subsp. *meridae* is found in western Venezuela; subsp. *baptiste* occurs in central Ecuador; subsp. *catharina* occurs in northern Peru and subsp. *sapphiropygia* is found in central and southern Peru

HABITAT Open, humid habitats such as forest edges and glades, paramo scrub and grassland; 9,200–15,750 ft (2,800–4,800 m)

SIZE Length: 4⅞–5½ in (12.5–14 cm). Weight: 5.4–6.5 g

STATUS Least Concern

MALE

MALE

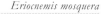

Eriocnemis mosquera

Golden-breasted Puffleg

This rather large puffleg is named for its diffuse, light yellow-orange breast patch; the plumage is otherwise shining green with a dark, glossy tail and wings. There are no contrasting colored areas on the forehead or vent, and the sexes are similar. The species' territoriality and aggression towards other hummingbirds are typical for the genus. It is most often seen feeding around low, dense bushes, usually hovering but sometimes clinging on with wings still beating. When it settles on a perch, it usually holds its wings raised for a moment. It selects a sheltered spot for its nest, with overhanging rocks or large leaves to protect it from heavy rainfall. The nest is a well-insulated cup made of moss and other plant material, fixed to a slim branch or dangling moss strand. This is quite a common bird, especially in some protected areas.

DISTRIBUTION Southwest and central Andes, from Colombia to northwest Ecuador

HABITAT Stunted and elfin forest, scrub, clearings, woodland edges; 3,950–11,800 ft (1,200–3,600 m)

SIZE Length: 4¾–5⅛ in (12–13 cm). Weight: 5.2–5.8 g

STATUS Least Concern

Emerald-bellied Puffleg

A small puffleg, this species has larger and more fluffy white leg puffs than any of its congeners. The plumage is bright, shining green, with a white breast patch and, in the male, a small iridescent patch of turquoise-green on the forehead. The rather short tail is light, shining blue-green. Two subspecies are recognized, subsp. *aline* and subsp. *dybowskii*, the latter with a smaller white breast patch. The species generally feeds at quite low levels within the forest strata, though not usually at or near ground level, and takes both nectar and insects readily. It is usually seen alone and may be aggressive to other hummingbirds at good feeding areas. The kinds of forests that seem best for it are those that have already been somewhat degraded, but such habitats are at risk of further destruction, and as a result the species is declining.

DISTRIBUTION Subsp. *aline* occurs in south-central and eastern Andes in Colombia into eastern Ecuador; subsp. *dybowskii* occurs in the eastern Andes of north and central Peru

HABITAT Cloud forest, smaller forest glades, generally less open habitats than other pufflegs; 7,550–9,200 ft (2,300–2,800 m)

SIZE Length: 3⅛–3½ in (8–9 cm). Weight: 4–4.5 g

STATUS Least Concern

MALE

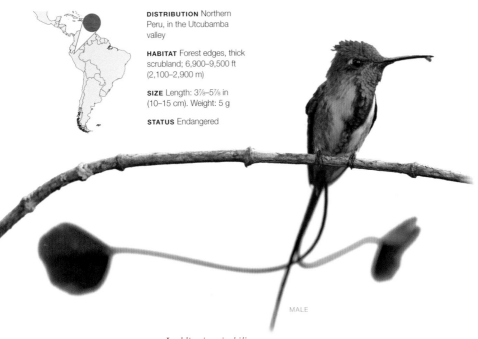

DISTRIBUTION Northern Peru, in the Utcubamba valley

HABITAT Forest edges, thick scrubland; 6,900–9,500 ft (2,100–2,900 m)

SIZE Length: 3⅞–5⅞ in (10–15 cm). Weight: 5 g

STATUS Endangered

MALE

Loddigesia mirabilis

Marvelous Spatuletail

This remarkable hummingbird is named for the male's unusual tail, which is formed by just two pairs of feathers. The central pair is slender and straight, while the outer pair is much longer, threadlike, outswept, and tipped with round violet "spatules." These dance and flutter around the bird during his dashing courtship flight. The female has a shorter, blunt tail, and lacks the male's blue crest and shining green throat. Males display in leks, competing for the females' attention. Occurring only in northern Peru, the species has a population of fewer than 1,500 individuals. Adult males are particularly scarce, partly because they are targeted by local hunters. About 100 acres (40 hectares) of the bird's habitat near Pomacocha village is now protected as a nature reserve, where it can be seen at feeding stations. Surveying to search for other populations is a priority, as is finding ways to reduce hunting.

Shining Sunbeam

This is a short-billed and heavy-bodied hummingbird with a mainly dark plumage, brown on the head and underparts, and sooty blackish brown on the wings, back, and tail. The rump is brightly iridescent and shades through light violet to gold and green. In subsp. *caumatonota*, the rump patch is almost entirely amethyst colored; this form also has darker body plumage than the nominate. The Shining Sunbeam is territorial (in the case of the males at least), defending rich clumps of flowers at any height in the strata. It clings to the flower to take nectar, and is also an avid flycatcher. It may nest at any time of year, building a solid cup-shaped nest on a branch or epiphyte well above ground level. It makes altitudinal movements between the seasons. Rather common and willing to use quite diverse habitat types, this species has a stable population.

DISTRIBUTION Subsp. *cupripennis* is found throughout the Andes of Colombia, Ecuador, and Peru; subsp. *caumatonota* occurs in central and south Peru

HABITAT Paramo, cloud forest, high ridges with scattered trees; 8,200–14,100 ft (2,500–4,300 m)

SIZE Length: 4¾–5⅛ in (12–13 cm). Weight: 6.9–8.1 g

STATUS Least Concern

MALE

White-tufted Sunbeam

This species is a typical sunbeam with a stocky outline, short bill, and mainly dark blackish-brown plumage, with a broad, glistening purple rump patch (a little less vivid in females). It is named for the scattered teardrop-shaped white feathers on its breast. Subsp. *regalis* is lighter in color than the nominate, and also has a chestnut rather than blackish-brown tail. The species is a territorial bird, often seen watching over its patch from a high perch, but it is subordinate to the Shining Sunbeam and keeps to lower levels where the two species occur together. It clings to flowers to feed, and gracefully pursues insects on the wing. It may undertake short seasonal movements to lower elevations, but can tolerate brief spells of severe weather. It has a restricted distribution only in Peru and is thought to be declining, but it has the advantage of occurring in protected areas, notably the Machu Picchu UNESCO World Heritage Site.

DISTRIBUTION Subsp. *castelnaudii* occurs in the Andes of south-central Peru; subsp. *regalis* occurs in the Andes of central Peru

HABITAT Patchy montane forest and forest edges; 11,500–15,100 ft (3,500–4,600 m)

SIZE Length: 4⅜–4¾ in (11–12 cm). Weight: 7–8.5 g

STATUS Least Concern

MALE

Purple-backed Sunbeam

A medium-sized, mainly dark hummingbird, the Purple-backed Sunbeam has white patches on its face, breast, and wing coverts, and the male's lower back and rump are bright, iridescent purple, shading into golden green. The female has a much reduced area of iridescence. The species is regarded by some authorities as a subspecies of the Shining Sunbeam. It is known to feed from the mistletoe *Tristerix longibrachteatus*, which parasitizes *Alnus* (alder) trees, and like other sunbeams is probably territorial. It is a little-known hummingbird with an extremely limited distribution in northern Peru, although a 2006 survey found it was more numerous than previously thought, leading to a change in status classification from Critically Endangered to Endangered. The population numbers no more than 4,000 individuals. Its known range is currently unprotected, and establishing a nature reserve at the site is a priority. Further local surveys and taxonomic work are also needed.

MALE

DISTRIBUTION Eastern Andean slopes of northern Peru

HABITAT Slopes with scattered vegetation, including *Alnus* trees; 9,850–10,500 ft (3,000–3,200 m)

SIZE Length: 4¾–5⅛ in (12–13 cm). Weight: 7.3–8.3 g

STATUS Endangered

Black-hooded Sunbeam

A very striking, medium-sized hummingbird with a proportionately large head and needlelike bill, this species is mostly velvety brownish black, but has a glistening turquoise-blue lower back and rump (the extent of the iridescence varies considerably). It also shows a variable white breast patch. The tail is rufous, tipped darker. The male's dark plumage is blacker than the female's and his iridescence is typically more extensive. This hummingbird is one of the most sought-after species in Bolivia for birdwatchers visiting the country, and occurs in several protected areas, including the Carrasco, Cotapata, and Amboró national parks. It feeds from *Barnadesia* and *Berberis* flowers and other bushy plants, invariably perching on, or clinging to, the plant as it takes nectar. Although it has a restricted range, the species seems quite adaptable in terms of habitat use and its population is believed to be stable.

DISTRIBUTION Cordillera Real, Bolivia

HABITAT Shrubby areas on the treeline, humid montane forest, primarily at the upper end of the altitudinal range; 5,900–11,500 ft (1,800–3,500 m)

SIZE Length: 5⅛ in (13 cm). Weight: 7–8.5 g

STATUS Least Concern

MALE

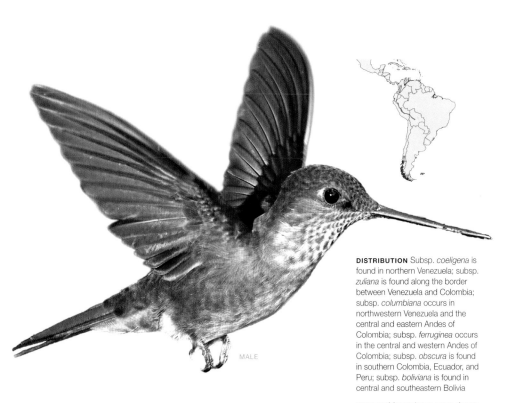

MALE

DISTRIBUTION Subsp. *coeligena* is found in northern Venezuela; subsp. *zuliana* is found along the border between Venezuela and Colombia; subsp. *columbiana* occurs in northwestern Venezuela and the central and eastern Andes of Colombia; subsp. *ferruginea* occurs in the central and western Andes of Colombia; subsp. *obscura* is found in southern Colombia, Ecuador, and Peru; subsp. *boliviana* is found in central and southeastern Bolivia

HABITAT Mountainous areas, forest edges, open areas, plantations; 4,600–8,550 ft (1,400–2,600 m)

SIZE Length: 5⅛–5⅞ in (13–15 cm). Weight: 6–7.5 g

STATUS Least Concern

Coeligena coeligena

Bronzy Inca

The Bronzy Inca is one of the duller members of the genus *Coeligena*, with a dark bronze-brown head and upperparts, a speckled white and gray breast, and a green rump. It has a long, straight black bill (slightly longer in the female) with yellow towards the base of the lower mandible, and small white spots behind each eye. The bronze tail feathers are forked. Subsp. *zuliana* is slightly greener, with less bronze in the head and upperparts; subsp. *columbiana* is the smallest subspecies and is a darker green; subsp.

ferruginea has a browner breast and underparts; subsp. *obscura* is much darker, with a grayer throat and larger dark spots on its breast; and subsp. *boliviana* has more purple-black tail feathers. The species is a trapliner, feeding on the flower nectar of many different species; it also catches insects. It builds a cup-like nest from plant fibers and moss, hidden in shrubs. The current global population is unknown, although it appears to be stable, and the species is regarded as fairly common throughout its range.

Black Inca

The Black Inca is purple-black above and underneath, with a shining blue-green throat patch and green rump. It also has white collar patches, shining blue shoulder patches, and small white spots behind each eye. The long, straight bill is black and the legs pinkish. Its black tail feathers are forked, and the undertail coverts have white edges. The female has a slightly longer bill and smaller shoulder patches. The species is a trapliner, feeding at all levels on the nectar of several plants—often with long, pendant flowers—including *Aetanthus*, *Fuchsia Macrocarpaea*, and *Psammisia falcata*; it also catches insects and small spiders. Its breeding habits are not recorded. There are thought to be between 3,500 and 15,000 individuals, with the species in moderate decline.

MALE

DISTRIBUTION North-central Colombia

HABITAT Mountainous areas, forest, open areas, riversides; 3,300–9,850 ft (1,000–3,000 m)

SIZE Length: 5⅛–5½ in (13–14 cm). Weight: 6–7 g

STATUS Vulnerable

Brown Inca

The Brown Inca is very similar to the Bronzy Inca, which it replaces below elevations of about 5,900 feet (1,800 m). It has a brown head and upperparts (which can appear reddish bronze), with small white spots behind each eye, a dull brown breast and underparts, and a green rump. Most distinctively, the bird has a shining purple throat patch with white collar patches. Its bronze tail feathers are forked. The black bill is long and straight, slightly longer in the female, which also has a smaller throat patch and a less forked tail. The species is a trapliner, mostly feeding at low and mid-levels on the nectar of flowering plants such as *Fuchsia*, *Abutilon*, *Bomarea*, *Columnea*, Bromeliaceae, *Psammisia*, and *Macleania*, and also takes insects and small spiders from plants. It builds a cup-like nest from soft plant fibers and moss, in the branches of young trees. The current global population is unknown although it may be in decline. However, the species is regarded as common throughout its range.

DISTRIBUTION Western and southwestern Colombia to western Ecuador

HABITAT Mountainous areas, forest; 2,600–8,200 ft (800–2,500 m)

SIZE Length: 4⅜–5⅛ in (11–13 cm). Weight: 6–7 g

STATUS Least Concern

MALE

Collared Inca

The Collared Inca is divided into eight subspecies that vary mainly in their head and throat patterns. The nominate subsp. *torquata* has a black head with a dark purple patch on the crown, greenish-black upper- and underparts, a green rump, and a wide white collar that broadens under the throat. It has a long, straight black bill, and small white spots behind each eye.

The white tail feathers are forked, and the outer tail feathers have dark green tips. The female has a slightly longer bill, and lacks the purple crown spot. The species is a trapliner, feeding at all levels on the nectar of many different plants, including *Abutilon*, *Bomarea*, *Capanea*, *Columnea*, Bromeliaceae, and *Macrocarpaea*; it also catches insects. It builds a cup-like nest from plant material, beneath ferns on rocky cliffs. The current global population is unknown but appears stable, and the species is regarded as very common throughout its range.

MALE

DISTRIBUTION Subsp. *torquata* occurs in northwestern Venezuela to Colombia; subsp. *conradii* ranges from northwestern Venezuela, Colombia, and eastern Ecuador to northern Peru; subsp. *fulgidigula* occurs in Ecuador; subsp. *margaretae* occurs in northern Peru; subsp. *insectivora* occurs in central Peru; subsp. *eisenmanni* occurs in southern Peru; subsp. *omissa* is found in southern Peru; subsp. *inca* occurs in northern and central Bolivia

HABITAT Mountainous areas, forests, forest edges; 4,600–9,850 ft (1,400–3,000 m)

SIZE Length: 5½–5⅞ in (14–15 cm). Weight: 6.5–7.5 g

STATUS Least Concern

Rainbow Starfrontlet

Another variable species, this bird is primarily green on its face, breast, and upper back, and orange on its wings, belly, and tail. The many subspecies vary principally in the iridescent areas on the male's forehead and throat. The nominate subsp. *iris* has a purple throat patch and a crown that shades from green through yellow to blue. Of the others, subsp. *hesperus* has a mainly gold and blue forehead; subsp. *aurora* has a light green forehead and light blue throat; subsp. *fulgidiceps* resembles subsp. *iris* but has a smaller throat patch; subsp. *flagrans* is also like subsp. *iris* but with a reddish rear crown; and subsp. *eva* is paler with a mostly yellow forehead patch. Willing to use modified habitats, the species is a trapliner that visits an extensive range of flowers. It also hawks and gleans insects and is common in some parts of its small range.

DISTRIBUTION Subsp. *iris* occurs from south Ecuador to north Peru; subsp. *hesperus* occurs in southwest Ecuador; subsp. *aurora* occurs in north Peru; subsp. *fulgidiceps* occurs in north Peru east of Marañón province; subsp. *flagrans* occurs in northwest Peru; subsp. *eva* occurs in north Peru west of the Marañón River on eastern slopes of the Andes

HABITAT Forest edges, gardens, riverside scrubland; 5,600–10,850 ft (1,700–3,300 m)

SIZE Length: 4⅞–5⅞ in (12.5–15 cm). Weight: 8–9 g

STATUS Least Concern

MALE

Coeligena phalerata

White-tailed Starfrontlet

This is a medium-sized hummingbird with a long, straight bill. The male is bright, shining green with a blue crown and throat, and a white lower belly, vent, and tail. The female is more olive green, with orange-buff underparts and a green crown and tail. Males are more often found within the forest interior, and females on forest borders, though there is considerable overlap. The species primarily visits low-growing flowers and is fairly territorial, more so than other *Coeligena* species. However, it sometimes also feeds by traplining, depending on availability and distribution of good nectar plants, and supplements its diet with insects caught in flight and picked from foliage. Although not yet considered threatened, it is declining. It has a very small geographic range and its habitat is under severe threat of deforestation (despite nominal protection).

MALE

DISTRIBUTION Sierra Nevada de Santa Marta, northeast Colombia

HABITAT Humid and wet forest and forest edges; 4,600–12,150 ft (1,400–3,700 m)

SIZE Length: 5½ in (14 cm). Weight: unknown

STATUS Least Concern

Dusky Starfrontlet

This is a newly recognized species, which was until recently considered to be a subspecies of the Golden-bellied Starfrontlet. The male is almost black, with a metallic gloss that shades to yellow-gold on the rump, and he has a bright blue chin patch and iridescent green forehead patch. Females lack these adornments and are more olive green, with yellow on the throat, belly, and rump. This bird, whose biology and ecology are still virtually unknown, has an estimated population of just 250 individuals, occupying two precariously tiny patches of forest in northwest Colombia. Very little other suitable habitat exists anywhere near the known sites. In 2005, a nature reserve to protect and study the species was established at the Páramo de Frontino, but the site is still poorly protected and is under high pressure from ore mining and human settlement. The local political instability causes further difficulties for any conservation workers.

DISTRIBUTION Two very small areas of forest—Páramo de Frontino and Farallones del Citará—in northwest Colombia

HABITAT Elfin forest and paramo, adjacent humid forest; around 11,500 ft (3,500 m)

SIZE Length: 5½ in (14 cm). Weight: 6.5–7 g

STATUS Critically Endangered

MALE

Buff-winged Starfrontlet

This dark green starfrontlet is distinguished from the other species in its genus by having cinnamon-colored tertials. Otherwise, the male is shining green below and blackish above, with a blue throat patch and green forehead patch, and the female is lighter with an orange-buff throat. Subsp. *albimaculata* differs from the nominate by having white tertials. A trapliner that often feeds while perched, this bird takes nectar at many flowers with a long corolla,

including species of *Bomarea*, *Cavendishia*, *Fuchsia*, *Nasa*, and *Tropaeolum*, and also catches insects. The nest is a large cup of soft plant material built on a branch fork. When not breeding, the species undertakes altitudinal movements. It is common in some areas and a very regular visitor to feeders in the Yanacocha Reserve near Quito, although the patchy distribution within its range suggests that it is not especially adaptable.

DISTRIBUTION Subsp. *lutetiae* is found in the Central Andes through Colombia and Ecuador (on the eastern slopes) into northern Peru; subsp. *albimaculata* is found on the western slopes of the Andes in northwest Ecuador.

HABITAT Cloud forest, *Polylepis* woodland; 8,550–11,800 ft (2,600–3,600 m)

SIZE Length: 5½ in (14 cm). Weight: 6.5–7 g

STATUS Least Concern

MALE

MALE

Coeligena bonapartei

Golden-bellied Starfrontlet

This starfrontlet is mainly shining green, shading to golden on the rump and belly, with dark wings. It often looks very dark apart from the golden areas. In addition, the male has a blue throat patch and vividly iridescent green forehead, these standing out against the blackish crown, nape, and upper cheeks. There are three subspecies: subsp. *bonapartei*; subsp. *consita*, in which the male has an orange band across the secondaries; and subsp. *eos*, which has more rufous secondaries. A quick-darting species, this is mainly a trapline feeder, visiting many flower types at both low and high levels; it also takes insects from foliage and in midair. The species breeds in the first half of the year and moves downslope when not breeding. Deforestation (including illegal logging) in its habitat is causing the species to decline, though at present it is still fairly common as it seems able to cope with some habitat degradation.

DISTRIBUTION Subsp. *bonapartei* occurs in the eastern Andes in Colombia; subsp. *consita* occurs in the Colombia–Venezuela border; subsp. *eos* is found in Venezuela

HABITAT Cloud forest, dwarf forest, patchy scrub; 4,600–10,500 ft (1,400–3,200 m)

SIZE Length: 5½ in (14 cm). Weight: 6.5–7 g

STATUS Least Concern

Blue-throated Starfrontlet

A very dark green hummingbird, slightly
smaller than other starfrontlets, this species
reveals patches of spectacular multicolored
iridescence in strong light. The male shows a
green forehead patch, violet-blue throat patch,
lighter blue rump, and sequin-like, shining
pink feathers on the belly. Females are quite
markedly smaller and lighter green, with
speckled orange underparts and a little violet-
blue sheen on the rump. There are two
subspecies, subsp. *helianthea* and the darker
subsp. *tamai*. This is a trapline feeder, moving
along its route at a low level in the vegetation,
gleaning insects as it goes, and sometimes
joining mixed flocks. Although long-billed,
it cannot reach all flowers' nectaries; it accesses
nectar from long *Passiflora mixta* corollas
through holes made in their bases by White-
sided Flowerpiercers (*Diglossa albilatera*).
This is an adaptable and often common
species, able to do well in somewhat
modified and degraded habitats.

FEMALE

MALE

DISTRIBUTION Subsp. *helianthea*
occurs in the north and east Andes
in Colombia; subsp. *tamai* occurs in
Páramo de Tamá, Táchira, western
Venezuela

HABITAT Cloud forest, dwarf forest,
scrub, well-planted gardens,
woodland edges; 6,250–10,850 ft
(1,900–3,300m)

SIZE Length: 5⅛ in (13 cm).
Weight: 6–7.5 g

STATUS Least Concern

MALE

Lafresnaya lafresnayi

Mountain Velvetbreast

A smallish hummingbird with a long, markedly downcurved bill, this species is represented across its range by seven subspecies. All share a generally dark green color scheme with a pale-sided tail. The breast and belly is black in males and white in females. Subsp. *liriope* differs from the nominate subsp. *lafresnayi* in having more green in the tail; subsp. *greenewalti* is paler and yellower overall; subsp. *longirostris* has a stronger buff tinge to its tail; and subspp. *saul, orestes,* and *rectirostris* have pure white outer tail feathers. Males are territorial, while females trapline; both visit many flower types and also catch insects. The female builds a large, sturdy cup nest under a sheltering overhang or on a tree trunk, and in it lays two eggs, which she incubates for 16 to 19 days. The chicks fledge at 23 to 26 days. This species has a wide range and a stable population.

DISTRIBUTION Subsp.
lafresnayi occurs in the far west
of Venezuela and eastern and
central Andean slopes in
Colombia; subsp. *liriope* occurs
in northern Colombia; subsp.
greenewalti occurs in west
Venezuela; subsp. *longirostris*
is found in central Colombia;
subsp. *saul* occurs in the
Andes of southwest Colombia,
Ecuador, and northern Peru;
subsp. *rectirostris* occurs in
central and southern Peru;
subsp. *orestes* is found in
northern Peru

HABITAT Montane forest edges,
humid scrub; 6,250–11,150 ft
(1,900–3,400 m)

SIZE Length: 4½–4¾ in
(11.5–12 cm). Weight: 4.5–6.5 g

STATUS Least Concern

FEMALE

Sword-billed Hummingbird

The extraordinary bill of this species makes up about half its total length—proportionately, it is the longest-billed bird in the world and must perch with its head tilted upwards to avoid overbalancing. The plumage is shining olive green, shading to brownish on the crown; the female has pale underparts with heavy green spotting and both sexes have a dark tail and wings. This species specializes in flowers with very long, pendent corollas, which it accesses by hovering or perching below them. It also hawks for insects, trapping flies in its open mouth in the manner of a swift or nightjar rather than snapping at them with the bill. Fairly common and very widespread, it is a much sought-after species for visiting birdwatchers and can be seen in various protected areas such as Munchique National Natural Park in Colombia and Pasochoa Forest Reserve in Ecuador.

DISTRIBUTION Along the Andes from west Venezuela through Colombia, Ecuador, and Peru to northeast Bolivia

HABITAT Humid montane forest, forest clearings and edges; 5,600–11,500 ft (1,700–3,500 m)

SIZE Length: 6¾–9 in (17–23 cm). Weight: 12–15 g

STATUS Least Concern

MALE

Great Sapphirewing

A large, broad-tailed hummingbird, this species is named for the male's shining blue wings. His plumage is otherwise glistening green, while the smaller female has reduced blue iridescence in the wings and a light orange throat, breast, and belly. There are three subspecies: subsp. *cyanopterus*; the bluer subsp. *caeruleus*; and the generally lighter-colored subsp. *peruvianus*. The species is territorial but will join mixed-species flocks at times. In flight, it shows noticeably slow, somewhat bat-like wingbeats. It often clings to flowers when feeding, and catches insects in flight. It rests on high, exposed branches early in the morning. The cup nest is built under sheltering vegetation, often suspended from tree roots. Incubation of the two eggs takes 16 to 18 days, and the female feeds the young in the nest for a further 26 to 29 days until they fledge. Common in some areas, the species numbers appear stable.

FEMALE

MALE

DISTRIBUTION Subsp. *cyanopterus* occurs in the eastern Andes of north-central Colombia; subsp. *caeruleus* occurs in the central and extreme southwest Andes in Colombia; subsp. *peruvianus* occurs in Ecuador, Peru, and northern Bolivia

HABITAT Humid open cloud forest and elfin forest, forest edges, scrub, wet grassland; 8,550–12,150 ft (2,600–3,700 m)

SIZE Length: 6¼–7⅞ in (16–20 cm). Weight: 9–11 g

STATUS Least Concern

Buff-tailed Coronet

This species is quite similar to the Chestnut-breasted Coronet, but has spotted green rather than chestnut underparts. The tail sides are light buff and the wings dark. Subsp. *flavescens* has a patch of dark chestnut on the inner secondaries, while subsp. *tinochlora* has a tinge of bronze on the buff tail sides. This coronet, like related species, feeds at all heights from flowers like *Erythrina, Inga,* and bromeliads, and is territorial around clumps of flowers. It usually clings to flowers to feed, with wings held in a raised posture. It also makes long hawking flights over clearings in pursuit of insects, and sometimes clings to tree bark and searches for arthropods, or takes sap from woodpecker holes. Across its range it is common in some areas and scarce in others—overall, its population trend is not clear and further studies are needed to clarify this.

DISTRIBUTION Subsp. *flavescens* is found in the Andes of northwest Venezuela and Colombia; subsp. *tinochlora* occurs in southwest Colombia and Ecuador

HABITAT Wet montane forest, cloud forest, dwarf forest, scrub; 6,550–11,500 ft (2,000–3,500 m)

SIZE Length: 4⅜–4¾ in (11–12 cm). Weight: 8–8.5 g

STATUS Least Concern

MALE

MALE

DISTRIBUTION The Andes through far southeast Colombia, Ecuador, and Peru

HABITAT Humid pre-montane forest interior; 3,950–9,850 ft (1,200–3,000 m)

SIZE Length: 4½–4¾ in (11.5–12 cm). Weight: 7–7.5 g

STATUS Least Concern

Boissonneaua matthewsii

Chestnut-breasted Coronet

This is a sturdy and relatively short-billed hummingbird with a slightly notched tail. It has shining green upperparts, a bright, glittering green throat, and a rich, warm chestnut breast and belly. The undersides of the wings and the tail sides are also chestnut and there is a small chestnut patch on the innermost secondaries; the wings are otherwise dark. When alighting on a perch, the bird holds its wings raised above its back momentarily.

Like other coronets it feeds at all heights in the forest strata. It sometimes clings to a flower to feed rather than hovering, and is territorial and aggressive. It is usually observed alone but on occasion several may be seen at flowering trees. It builds a sturdy nest from moss and lichen. The species' general ecology and population trend are not well known, but it appears to be fairly common over its extensive range and shows no marked geographical variation.

Velvet-purple Coronet

This spectacular medium-sized coronet has a black head, primaries, and central tail feathers, white tail sides, and small white leg puffs. The throat patch, crown, back, and belly are shining violet-blue, while the rest of the wings, back, and rump are more turquoise-green. The female has a black forehead and less extensive iridescence. The species is uncommon and may be difficult to observe, especially as it feeds up to tree-top level, taking both nectar and insects. The male holds a feeding territory. In courtship, the male displays by circling the female, singing a sequence of shrill and scratchy notes, and hovering with the tail fanned. The female constructs her cup-shaped nest, woven from moss and lichen, on a horizontal branch. The species is a regular and common visitor to feeders at sites between Nanegalito and Mindo in northwestern Ecuador.

DISTRIBUTION Pacific slopes of the western Andes in southwest Colombia and northwest Ecuador

HABITAT Damp, dense forest, forest edges and clearings; 1,150–7,200 ft (350–2,200 m)

SIZE Length: 4⅜–4¾ in (11–12 cm). Weight: 8–8.5 g

STATUS Least Concern

MALE

Booted Racket-tail

The male of this species has greatly extended
central tail feathers, that are sometimes crossed
over, with long, bare shafts tipped with shining
blue discs. The shape and length of these
feathers vary between the eight described
subspecies. Both sexes have fluffy leg puffs,
which are white in some subspecies and orange
in others. Males are otherwise shining green,
and females are short-tailed with green-spotted
white underparts. This bird feeds in the forest
interior at all heights. It is non-territorial and
tolerates others of its species in close proximity
as it takes nectar from *Palicourea*, *Inga*, and
similar plants. Courting males display their
contrasting leg puffs and audibly snap the
elongated tail feathers in a series of hovers and
dives. The nest is a tiny cup built on a horizontal
twig high in a tree. This is a very common bird
over most of its range.

FEMALE

MALE

DISTRIBUTION Subsp. *underwoodii* is found in the eastern Andes of Colombia; subsp. *polystictus* occurs in north Venezuela; subsp. *discifer* is found in northwest Venezuela; *incommodus* occurs in the west and central Andes of Colombia; subsp. *melanantherus* occurs in the Andes of Ecuador; subsp. *peruanus* occurs in east Ecuador and northeast Peru; subsp. *annae* is found in central and south Peru; subsp. *addae* occurs in Bolivia

HABITAT Wet forest and secondary growth; 1,950–13,100 ft (600–4,000 m)

SIZE Length: 3–5⅞ in (7.5–15 cm). Weight: 2.5–3 g

STATUS Least Concern

White-tailed Hillstar

A large, stocky, long-billed hummingbird, this
species is rather a dark, mainly olive-green color,
with white tail sides and a large blue throat and
breast patch. The sexes look similar. Just two
subspecies are known: subsp. *bougueri*, which
has bright orange patches at the sides of the bill
base; and subsp. *leucura*, which lacks the orange
facial markings. The species is often found
around streams, where it hunts for insects over
the water, and it also visits flowering trees and
shrubs for nectar. The male defends clumps of
flowers at various heights in the strata. The
female builds a cup nest on a vertical branch,
well above ground level, and incubates the eggs
for 16 to 18 days. The chicks take 23 to 25 days
to fledge, by which time they resemble adults
apart from some buff fringes on the head
feathers. The species' population has not
yet been properly evaluated.

MALE (SUBSP. *BOUGUERI*)

MALE (SUBSP. *LEUCURA*)

DISTRIBUTION Subsp. *bougueri* occurs on the Pacific slope of the Andes in southwest Colombia and northwest Ecuador; subsp. *leucura* occurs on the eastern slope of the eastern Andes in south Colombia, east Ecuador, and northeast Peru

HABITAT Montane forest and forest edges, secondary growth, scrubby pasture; 5,250–9,200 ft (1,600–2,800 m)

SIZE Length: 5⅛–5½ in (13–14 cm). Weight: 8.5–9 g

STATUS Least Concern

Purple-bibbed Whitetip

This species is very similar to the Rufous-vented Whitetip but is a little smaller. The male's white tail spot is also proportionately larger, and it also has a narrow, shining violet throat patch. The population in Peru has sometimes been considered a separate subspecies, subsp. *intermedia* (variously classed as a subspecies of either Purple-bibbed or Rufous-vented). However, most modern taxonomies do not recognize subsp. *intermedia* and consider both whitetips to be monotypic. This species is usually seen alone, feeding from ground level to the forest canopy, from flowers such as *Inga* and *Cinchona*. It hovers to take nectar and also hawks for insects. It nests up to 10 feet (3 m) above ground, building a cup nest from fern filaments and moss within a shrub or among vines. It is fairly common and occurs in at least two significant protected areas.

FEMALE

MALE

DISTRIBUTION Pacific slopes of Western Colombia to northwest Ecuador

HABITAT Forests, forest edges; 2,300–5,250 ft (700–1,600 m)

SIZE Length: 3⅛–3½ in (8–9 cm). Weight: 4 g

STATUS Least Concern

Rufous-vented Whitetip

The two very similar *Urosticte* species, which are classified by some authorities as subspecific forms of the same species, take their English name from the white tips to the male's central tail feathers. These feathers are curved upwards, making the white area stand clear of the rest of the tail. The male is otherwise glistening green with a patchy whitish breast band and russet undertail. The female has white underparts marked with round, shining green spots and white-tipped outer tail feathers, and the tail is also less deeply notched than the male's. This montane species forages for nectar low down and at mid-height in the forest strata, visiting *Clusia*, *Palicourea*, and Bromeliaceae flowers. It builds a neat, mossy cup nest among hanging vines 6 to 13 feet (2 to 4 m) above ground. This species is declining, particularly in Ecuador, but does have strong populations in several protected areas in Colombia.

DISTRIBUTION Eastern slope of south-central Andes in Colombia and east Ecuador

HABITAT Humid but rather open forest and cloud forest; 5,250–7,850 ft (1,600–2,400 m)

SIZE Length: 3½–3⅞ in (9–10 cm). Weight: 4 g

STATUS Least Concern

MALE

Velvet-browed Brilliant

One of the smaller brilliants, this species has
the group's typical dark, shining green plumage
and characteristic tapering head shape. The
male's small throat patch is dark violet-blue;
the female has pale underparts marked with
large, shining green spots. A lively forest bird
with an agitated, chattering call, the species
feeds at both high epiphytes and low shrubs,
and it is common to see two or three together.
It mainly feeds by traplining but sometimes
shows spells of territoriality around a
particularly good resource. Often areas hold
mainly males or mainly females, suggesting
that at times the sexes have different ecologies.
The nest is saddle-shaped, built on a horizontal
branch. The species has a very patchy distribution
within its quite large range, though there are
some areas of suitable habitat that have not yet
been properly surveyed, and it seems to be an
altitudinal migrant.

DISTRIBUTION Subsp. *xanthogonys*
occurs in south and east Venezuela
and Guyana; subsp. *willardi* occurs
in southern Venezuela

HABITAT Forests, glades, forest
edges, scrub; 2,300–6,550 ft
(700–2,000 m)

SIZE Length: 3⅞–4⅜ in
(10–11 cm). Weight: 7 g

STATUS Least Concern

MALE

Black-throated Brilliant

This rather heavily built brilliant has a deeply forked tail. Males have a black lower face and chin, dark violet throat patch, green lower breast, and a blackish belly. The female's face is green rather than black. There are two subspecies: subsp. *schreibersii*; and subsp. *whitelyana*, the male of which has entirely black underparts apart from a violet throat patch. The bird feeds at quite high levels in the strata, taking nectar from epiphytes and flowering trees, and hawking for insects. Its behavior is otherwise very little known; although it has an extensive distribution, it is uncommon and hard to observe. It is, however, thought to be declining. Unfortunately, like most rainforest species it is at risk from the continued high rates of deforestation in Amazonia. The bird can, however, be seen in the protected Manú National Park in Peru.

DISTRIBUTION Subsp. *schreibersii* occurs in southeast Colombia, east Ecuador, and northeast Peru north of the Amazon River to the far northwest of Amazonian Brazil; subsp. *whitelyana* occurs in east Peru

HABITAT Humid tropical forest and scrub; 1,300–4,250 ft (400–1,300 m)

SIZE Length: 4¾ in (12 cm). Weight: 7–10 g

STATUS Least Concern

MALE

Gould's Jewelfront

Probably the most distinctive species in the genus *Heliodoxa*, this hummingbird has a shorter bill than most of its congeners and a more olive-toned plumage, with a glistening, light violet forehead patch in the male and a spotted throat in the female. Both sexes have a broad orange-buff breast band and some orange on the outer tail feathers. Gould's Jewelfront is a rather unobtrusive species that feeds in the understory and occasionally higher, hovering at flowering shrubs and taking insects in flight and from foliage. It is not strongly territorial but regularly raids territories held by smaller hummingbird species; it also traplines, and is attracted to mixed feeding flocks. Although found across a very large area, the species is quite uncommon in most parts of its range, and is undoubtedly declining due to the rate of deforestation in Amazonia. However, it does appear to accept somewhat modified habitats.

MALE

FEMALE

DISTRIBUTION Primarily Amazonia, from south Venezuela through Colombia, eastern Ecuador, and eastern Peru to north Bolivia and northwest Brazil

HABITAT Riverine and damp areas in humid forest; 500–1,950 ft (150–600 m)

SIZE Length: 4⅜–4¾ in (11–12 cm). Weight: 6.5 g

STATUS Least Concern

Fawn-breasted Brilliant

This medium-sized hummingbird has a dark green head, shining bronze-green upperparts, a pale orange breast with obscure green mottling, and otherwise pale underparts. It has a sparkling pinkish throat patch (absent in the female), white spots behind the eyes, and a medium-length, slightly downcurved black bill. The tail is reddish brown and slightly forked. Subsp. *aequatorialis* is more gold above, and has more green mottling underneath; and subsp. *cervinigularis* is slightly larger, and has a smaller, paler throat patch and less green mottling underneath. The species spends much of its time in low to mid-levels of the forest, but can be found in more open areas. It feeds on the nectar of many different plant species, including species of *Heliconia* in the understory and *Inga* and *Erythrina* higher up, as well as small insects. Very little is known about the breeding behavior of these birds. The current global population and population trend are unknown, and while numbers appear stable, the species is regarded as uncommon and patchily distributed through its range.

DISTRIBUTION Subsp. *rubinoides* occurs in Colombia; subsp. *aequatorialis* occurs in Colombia and Ecuador; subsp. *cervinigularis* occurs in Ecuador and Peru

HABITAT Mountainous areas, rainforest, forest edges, pastures, gardens; 2,600–6,550 ft (800–2,000 m)

SIZE Length: 3⅞–5⅛ in (10–13 cm). Weight: 7–9 g

STATUS Least Concern

MALE

Green-crowned Brilliant

The male of this medium-sized hummingbird has shining green upperparts, a dark green head and brilliant green forehead, and green underparts. It has a small purple throat patch, white spots behind the eyes, and a medium-length, straight black bill. The forked tail is dark blue. The female is whitish below, spotted with green, and has a white stripe below the eye. She lacks the throat patch and has a slightly longer, downcurved bill and a more shallowly forked tail with white tips on the outer feathers. Subsp. *henryi* is slightly larger than the nominate, with more brilliant green areas; and subsp. *jamersoni* has a shorter tail and much duller upperparts. The species spends much of its time in low to mid-levels of the forest, but can also be found in more open areas. It feeds on the nectar of many different plant species, including *Heliconia*, and on small insects. The cup-like nest is built from plant material on a downsloping branch, usually beneath a leaf. The current global population is unknown, although numbers do appear to be decreasing. The species is fairly common but is patchily distributed through its range.

MALE

FEMALE

DISTRIBUTION Subsp. *jacula* occurs from eastern Panama to north and central Colombia; subsp. *henryi* occurs in Costa Rica and western Panama; subsp. *jamersoni* occurs in southwestern Colombia and western Ecuador

HABITAT Mountainous areas, rainforest, forest edges; 2,300–6,550 ft (700–2,000 m)

SIZE Length: 3⅞–5½ in (10–14 cm). Weight: 7–10 g

STATUS Least Concern

FEMALE

Heliodoxa imperatrix

Empress Brilliant

This dramatic, large hummingbird has a long tail that is much more deeply forked than those of any of the other brilliants. The male is a deep, brilliant green with a golden belly and a small, shining violet throat patch; the female has a green-spotted pale underside, as is typical of the genus. This species is a solitary forager, flying with slow and steady wing beats. It feeds at low to mid-levels in the forest strata, seeking out hanging flowers with long corollas and hovering below them to access the nectar. It also hawks for insects and gleans them from foliage. Females are usually seen at lower levels than males. Few nests have been observed, but one example was a cup of soft kapok fibers built on top of a palm frond. The species has a small range but its habitat is largely intact and it is fairly common.

DISTRIBUTION Pacific slopes of west-central Colombia to northwest Ecuador

HABITAT Cloud forest and other very humid forest types, including secondary growth; 1,300–6,550 ft (400–2,000 m)

SIZE Length: 4¾–6¾ in (12–17 cm). Weight: 8.5–9.5 g

STATUS Least Concern

MALE

Violet-fronted Brilliant

The male of this medium-sized, dark species has a glistening, dark blue forehead patch and is otherwise shining green, more bronzy on the underside, with a dark tail and wings. The female has a pale throat, breast, and belly, marked with large, dark green spots. There are four subspecies: subsp. *leadbeateri*; subsp. *parvula*, which has a more violet forehead; subsp. *sagitta*, which is generally more blue-toned; and subsp. *otero*, which is less bronze-toned. It feeds at all heights in the forest strata, generally on its own. It sometimes holds a small territory but also traplines, and will come to hummingbird feeders from time to time if they are available on its traplining circuit. Its diet contains a high proportion of insects. This species is not well studied and at present its population trend is not known.

FEMALE

DISTRIBUTION Subsp. *leadbeateri* occurs in coastal mountains in north Venezuela; subsp. *parvula* occurs in northwest Venezuela through north, central, and south Colombia; subsp. *sagitta* occurs in east Ecuador and northern Peru; subsp. *otero* occurs in central and south Peru to north Bolivia

HABITAT Rainforest, cloud forest, forest edges; occasionally scrub, plantations, and woodland; 1,300–7,850 ft (400–2,400 m)

SIZE Length: 4⅜–5⅛ in (11–13 cm). Weight: 6.5–8.5 g

STATUS Least Concern

Brazilian Ruby

This fairly small hummingbird, the only species in its genus, has a mid-length, straight bill. It is primarily shining mid-green with rufous in the wings and tail, and shows a prominent white spot behind the eye. The male has a deep red throat patch, while the female has an entirely cinnamon-orange rump and underparts. There is a fairly common melanistic form, which is glossy blackish green with a dark violet throat patch. This common species is territorial in both sexes, and its extensive diet includes nectar from eucalyptus and banana flowers. It makes a soft cup nest on a horizontal branch well above ground, camouflaging the outside of the nest with lichen. Incubation takes 15 to 16 days and fledging about 25 days. Fledged young have female-like plumage. The species is common across much of its range and readily accepts non-natural habitats.

MALE

DISTRIBUTION Eastern Brazil along the coast into Rio Grande do Sul and inland through Goiás

HABITAT Forest, scrubland, gardens, parks, banana plantations; 0–3,300 ft (0–1,000 m)

SIZE Length: 4⅜–4½ in (11–11.5 cm). Weight: 6–9 g

STATUS Least Concern

Giant Hummingbird

This is the largest hummingbird species, unmistakable with its slow-flapping, sometimes gliding flight. It is dull olive green above with a slight sheen, and the rump is white. The male is clear cinnamon below. The female's underside is duller and mottled with light gray; her rump patch is also mottled gray. There are two subspecies. The northern subsp. *peruviana* is larger than the nominate subsp. *gigas*, and the male's orange coloration is brighter and extends to most of the face. The most southerly breeding populations migrate north in the southern winter. Giant Hummingbirds feed from large flower clusters, usually perching, and favor *Agave* and *Puya* flowers and various cacti. They are aggressive and territorial. The nest is small given the bird's size, made of moss, animal hair, and spiders' webs; occasionally, the clutch comprises just one egg. This species is quite common, with stable numbers.

DISTRIBUTION Subsp. *gigas* occurs in central and southern Chile and central-western Argentina; subsp. *peruviana* occurs in the Andes in southwest Colombia through Ecuador, Peru, and Bolivia to north Chile

HABITAT Dry, open habitats with low-lying vegetation; 0–15,750 ft (0–4,800 m)

SIZE Length: 7⅞–8¾ in (20–22 cm).
Weight: 18.5–20 g

STATUS Least Concern

FEMALE

Canivet's Emerald

This species, also known as Fork-tailed Emerald, is sometimes grouped with the Blue-tailed Emerald; its taxonomy is still far from resolved. Until the late 1990s it included the Cozumel Emerald, and subsp. *salvini* is also frequently split as a separate species. This bird is a typical shining green emerald (with a pale grayish underside in the female) and has a moderately long, forked tail. The male's bill is red with a dark tip. Subsp. *osberti* has a shorter tail than the nominate, and subsp. *salvini* has a darker bill. The species feeds low down and visits mainly small flowers that have short corollas, wagging and fanning its tail as it feeds. Males sing from low perches in bushes that adjoin open areas, and use a diving display both in aggressive encounters with other males and when courting females. This is a common and adaptable species.

DISTRIBUTION Subsp. *canivetii* ranges from southeast Mexico south to north Belize and north Guatemala, and to the Bay Islands of Honduras; subsp. *osberti* is found on the Pacific slope from south Mexico to Honduras; subsp. *salvini* is found on the Pacific slope through Nicaragua to northwest Costa Rica

HABITAT Most lightly vegetated habitat types, including savanna, scrub, light woodland, gardens, and plantations; 3,300–8,550 ft (1,000–2,600 m)

SIZE Length: 2½–3⅜ in (6.5–8.5 cm). Weight: 3–3.5 g

STATUS Least Concern

MALE

Chlorostilbon assimilis

Garden Emerald

Until recently this species was considered to be a subspecies of the Blue-tailed Emerald. It is a small hummingbird with a short, straight bill and shallow tail fork. The male is shining olive green with a darker tail and wings, while the female has a white throat, breast, and belly. The species is usually a low-level feeder but will go into the canopy to visit favorite trees when they are in flower. Both males and females are trapliners, patrolling a circuit and visiting flowers along the way—the delay between visits to each flower allows them time to replenish their nectar stores. The female builds a nest in thick cover, choosing an angled but not vertical branch. This is a common bird across much of its range, readily taking to modified habitats, although its population trend has not been clearly established.

DISTRIBUTION Southwest Costa Rica and the Pacific coast of Panama, including some islands

HABITAT Woodland edges and clearings, gardens; 3,300–8.550 ft (1,000–2,600 m)

SIZE Length: 2½–3⅜ in (6.5–8.5 cm). Weight: 3–3.5 g

STATUS Least Concern

Western Emerald

The Western Emerald—also known as the West Andean Emerald—is closely related to the Blue-tailed Emerald, and is sometimes regarded as a subspecies of the latter. It is greenish bronze above, with a shining goldish-green head, a bright green forehead and breast, and green underparts. It has a short, straight black bill and a bright, shining blue tail with a pronounced fork. The female is paler green above and gray below, and has a distinctive dark face mask with a small white line above. A trapliner, the Western Emerald usually feeds at low levels on the nectar of *Acnistus*, *Impatiens*, *Lantana*, Ericaceae, and *Salvia* flowers, among many others, as well as on small insects. The bird builds a cup-like nest from plant material. The current global population and population trend are unknown but as far as can be determined seem stable, and the species is regarded as fairly common throughout its range.

DISTRIBUTION Western Colombia and Ecuador

HABITAT Mountainous areas, forests, forest edges; 0–9,850 ft (0–3,000 m)

SIZE Length: 2⅜–3½ in (6–9 cm). Weight: 3–4 g

STATUS Least Concern

MALE

Red-billed Emerald

The Red-billed Emerald is closely related to the Blue-tailed Emerald, and is sometimes regarded as a subspecies of the latter. Like the Blue-tailed, it is greenish bronze above, with a shining goldish-green head, bright green forehead and breast, and green underparts. Its bright, shining blue tail has a pronounced fork. The bill is short and straight, with a red lower mandible. The female is very similar to the Blue-tailed female, with an all-black bill, and also has a distinctive dark face mask with a small white line above. The species is a trapliner, usually feeding at low levels on the nectar of small, primarily insect-pollinated flowers; it also catches small insects. The bird builds a cup-like nest from plant material. The current global population and population trend are unknown but seem stable, although the species is regarded as uncommon and patchily distributed in its range.

MALE

DISTRIBUTION Subsp. *gibsoni* occurs in central Colombia; subsp. *nitens* occurs in northeast Colombia and northwest Venezuela; subsp. *chrysogaster* occurs in northern Colombia and northwest Venezuela

HABITAT Mountainous areas, forests, forest edges, open areas, arid areas, shrublands, scrublands, parks, gardens; 0–9,850 ft (0–3,000 m)

SIZE Length: 2⅜–3½ in (6–9 cm). Weight: 3–4 g

STATUS Least Concern

Blue-tailed Emerald

The Blue-tailed Emerald has been divided into seven subspecies (see "Distribution"), mainly differing in color, particularly of the bill, and in the shape of the bright blue forked tail. The closely related Western and Red-billed Emeralds are sometimes regarded as subspecies of the Blue-tailed. The nominate subspecies is greenish bronze above, with a shining goldish-green head, bright green forehead and breast, and green underparts. It has a short, straight bill (black in females of all subspecies). The female is paler green above and gray underneath, with small white spots behind each eye and dark ear coverts. A trapliner, the Blue-tailed Emerald usually feeds at low levels on the nectar of many small, primarily insect-pollinated flowers, as well as other plants, and also catches small insects. The bird builds a cup-like nest. The current global population is unknown but seem stable, and the species is regarded as common.

DISTRIBUTION Subsp. *mellisugus* occurs in Suriname, French Guiana, and Brazil; subsp. *caribaeus* occurs in eastern Colombia, northeastern Venezuela and the islands of Curuçao, Aruba, Bonaire, Trinidad, and Margarita; subsp. *duidae* occurs in southern Venezuela; subsp. *subfurcatus* occurs in eastern and southern Venezuela, Guyana, and northwestern Brazil; subsp. *phoeopygus* occurs in eastern Peru; subsp. *napensis* occurs in northeastern Peru; subsp. *peruanus* occurs in southeastsern Peru and northern Bolivia

HABITAT Savanna, woodland, plantations, gardens, fields, hedgerows; 0–9,850 ft (0–3,000 m)

SIZE Length: 2⅜–3½ in (6–9 cm). Weight: 3–4 g

STATUS Least Concern

MALE

Cuban Emerald

The Cuban Emerald is both common and widespread on its namesake island and on the three largest islands of the Bahamas: Grand Bahama, Abaco, and Andros. Like many hummingbirds, both sexes will gather in noisy, fractious mobs around particularly rewarding nectar sources. In the Bahamas, they compete with Bananaquits (*Coereba flaveola*), wintering wood-warblers, and other songbirds for access to the dense yellow flower clusters of the Bahama Agave (*Agave braceana*). Females are similar in appearance to other female emeralds, though their dark bronze tails are very long and deeply forked, with narrow gray tips on the outer feathers. Gregarious nesting has been reported in Cuba, with six active nests averaging only a foot (30 cm) apart. Though Cuban Emeralds are considered sedentary, vagrants have been reported from New Providence Island in the Bahamas (former home of the now extinct Brace's Emerald), Turks and Caicos Islands, and Florida.

DISTRIBUTION Cuba, including Isla de la Juventud (formerly Isle of Pines) and some cays, and the Bahamas (Grand Bahama, Abaco, and Andros)

HABITAT Open forest, pine woodlands, coastal scrub, swamp edges, plantations, parks, gardens; 0–650 ft (0–200 m)

SIZE Length: 3½–4⅜ in (9–11 cm). Weight: 3–4.5 g

STATUS Least Concern

MALE

Glittering-bellied Emerald

The male Glittering-bellied Emerald is dark, shining goldish green above, with darker green on the head and forehead, a shining blue-green throat and breast, and green underneath. The shining blue tail has a slight fork. The short, straight bill is orange-red, turning black towards its tip. The female is similar, but is paler green above and the throat and breast are pale gray, turning darker gray-brown underneath; it has more black on the bill. Subsp. *pucherani* is slightly smaller than the nominate; subsp. *berlepschi* is less goldish green; and subsp. *igneus* has more sparkling orange underparts. The species is a trapliner, usually feeding at low levels on the nectar of a wide range of plants, including *Lantana camara*, *Duranta arboreus*, and *Citrus sinensis*; it also takes small insects. The bird builds a small cup-like nest from plant material. The current global population and population trend are unknown but appear stable, and the species is regarded as common throughout its range.

FEMALE

DISTRIBUTION Subsp. *lucidus* occurs in east Bolivia, Paraguay, and southwest Brazil; subsp. *pucherani* occurs in eastern Brazil; subsp. *berlepschi* occurs in southern Brazil, Uruguay, and northeastern Argentina; subsp. *igneus* occurs in northern Argentina

HABITAT Semiarid areas, savanna, scrubland, forest edges, parks, gardens; 0–11,500 ft (0–3,500 m)

SIZE Length: 2¾–4⅜ in (7–11 cm). Weight: 3–5 g

STATUS Least Concern

Green-tailed Emerald

This species is very similar to both the Narrow-tailed and Short-tailed emeralds, and is known to have hybridized with the latter. The male has entirely shining olive-green plumage apart from the darker flight feathers. The female's underparts are dull buffish white, with some green on the flanks. Like other emeralds this is primarily a trapline feeder and trespasser into other species' territories, but males have been seen to defend territories of their own around particularly rich flower clusters. The nest is a cup built saddle-style on a low, thin, horizontal branch, and is composed of a variety of plant material, including fluffy seeds, bark strips, moss, and leaf fragments. The incubation period is 14 days, with fledging taking place at 20 to 22 days. The species may undertake altitudinal movements, as its occurrence is markedly unpredictable. The population trend has not been established but birds seem to be fairly common.

FEMALE

DISTRIBUTION Northern Venezuela

HABITAT Rainforest and cloud-forest edges, secondary growth, gardens, plantations; 2,300–5,900 ft (700–1,800 m)

SIZE Length: 2½–3⅜ in (6.5–8.5 cm). Weight: 3–4 g

STATUS Least Concern

Short-tailed Emerald

Despite its name, this species is not noticeably shorter-tailed than other similar species, including the Green-tailed and Narrow-tailed emeralds. The male is entirely shining olive green apart from darker flight feathers, while the female has a white belly. As with other emeralds, the male is somewhat larger. There are two subspecies, subsp. *poortmani* and the slightly larger and more yellow-tinged subsp. *euchloris*. The species feeds primarily by traplining, following a route through more open areas and selecting low but non-hanging flowers like *Elleanthus* and *Phaseolus*. Its flight style is more insect-like than that of other emeralds, recalling a woodstar or coquette. The nest cup is a soft, springy structure made mainly from woolly plant seeds, the outside decorated with assorted plant fragments. Incubation takes 15 to 16 days and chicks fledge at 20 to 22 days. The species has a restricted range but is not uncommon and uses some modified habitats.

MALE

DISTRIBUTION Subsp. *poortmani* occurs in east Colombia and northwest Venezuela; subsp. *euchloris* occurs in central Colombia

HABITAT Humid forest and forest edges, woodland clearings, plantations, secondary growth, usually along streams; 350–7,850 ft (100–2,400 m)

SIZE Length: 2½–3⅜ in (6.5–8.5 cm). Weight: 3–4 g

STATUS Least Concern

Broad-billed Hummingbird

Despite its name, the bill of the broad-tailed, glitteringly iridescent Broad-billed Hummingbird is no broader than that of many of its close relatives. The blue of the male's throat and upper breast varies regionally from deep indigo to turquoise across the five subspecies, and is most extensive in the slightly smaller males of the Pacific lowlands of southern Mexico (subsp. *doubledayi*). Females are similar to females of the closely related *Chlorostilbon* emeralds, with gray underparts, prominent face markings, and white tips on the outer tail feathers. Northern populations are partially migratory, moving north and inland during the breeding season. The winter range has expanded with the increasing prevalence of feeders and winter-blooming hummingbird gardens in the southern and southwestern United States. Post-breeding wanderers have been recorded in the Great Lakes region, southeastern Canada, the Rocky Mountains, and the Pacific flyway.

MALE

DISTRIBUTION Subsp. *latirostris* occurs in east-central Mexico; subsp. *doubledayi* occurs in southern Mexico; subsp. *lawrencei* is restricted to the Marías Islands off the western coast of Mexico; subsp. *magicus* ranges from the southwestern United States to central-western Mexico; subsp. *propinquus* occurs in central Mexico

HABITAT Tropical deciduous forest, thorn scrub, foothill and canyon woodlands, riparian edges; 0–7,200 ft (0–2,200 m)

SIZE Length: 3⅛–3⅞ in (8–10 cm). Weight: 3–4 g

STATUS Least Concern

Blue-chinned Sapphire

This is the only species in the genus *Chlorestes*, although it is sometimes placed in *Chlorostilbon* with the emeralds and its similarity to those species is marked. It is a small hummingbird with mainly green plumage, a darker tail and wings, and, in the male, a diffuse blue throat patch. Unlike females of *Chlorostilbon*, the female's underside is green, with just a hint of white. Aside from the nominate there are two subspecies: subsp. *puruensis*, which has a smaller blue throat patch; and subsp. *obsoleta*, which has no blue on the throat. Males of this species are territorial and are often seen in gardens around clumps of flowers like *Hibiscus* and *Heliconia*, feeding 3 to 20 feet (1 to 6 m) above ground. Females make a small, deep cup nest made of lichen and other fine plant material. It is positioned low on a tree on a thin branch or root. This is a very widespread, common, and adaptable hummingbird, and there have even been occasional records of it as far afield as Tobago.

MALE

DISTRIBUTION Subsp. *notata* occurs from northeast Colombia across northeast Venezuela, Trinidad and Tobago, and the Guianas to east Brazil; subsp. *puruensis* occurs in northwest Brazil to southeast Colombia and northeast Peru on the upper Ucayali River; subsp. *obsoleta* occurs in northeast Peru on the lower Ucayali River

HABITAT All kinds of open and partly wooded countryside, including gardens and plantations; 0–1,650 ft (0–500 m)

SIZE Length: 2¾–3½ in (7–9 cm). Weight: 3–4 g

STATUS Least Concern

Violet-headed Hummingbird

This is a small, short-billed hummingbird with a slightly pointed crest. Both sexes are dark, shining green with a glossy violet-blue crown that tapers to a small point. The male also has a blue throat and the female is whitish below. There are three subspecies: subsp. *guimeti*, described above; subsp. *merrittii*, with a bluer rather than violet-toned crown; and subsp. *pallidiventris*, in which both sexes have paler underparts. This bird visits a range of trees, favoring small-flowered species, and its distribution shifts through the year according to the availability of these flowers. Males are sometimes territorial but only in the absence of larger hummingbirds, and both sexes hawk for insects around streams. Males form leks in the breeding season. The nest is a small sturdy cup on a dangling twig, often over water. The species is widespread, occurs in several protected areas, and is under no immediate threat, but it is declining and has a rather fragmented distribution.

DISTRIBUTION Subsp. *guimeti* occurs in north and west Venezuela, through east Colombia and Ecuador to north and northeast Peru; subsp. *merrittii* occurs from eastern Honduras to east Panama; subsp. *pallidiventris* occurs in east Peru and west-central Bolivia

HABITAT Tall humid forest, plantations, shaded gardens; 0–6,250 ft (0–1,900 m)

SIZE Length: 3–3⅜ in (7.5–8.5 cm). Weight: 2.5–3 g

STATUS Least Concern

MALE

MALE

Orthorhyncus cristatus

Antillean Crested Hummingbird

DISTRIBUTION Subsp. *cristatus* is found on Barbados; subsp. *exilis* ranges from eastern Puerto Rico through the Lesser Antilles to St. Lucia; subsp. *ornatus* is found on St. Vincent; subsp. *emigrans* occurs on Grenada and the Grenadines

HABITAT Gardens, parkland, plantations, woodland edges; 0–1,650 ft (0–500 m)

SIZE Length: 3⅛–3¾ in (8–9.5 cm). Weight: 3.5–4 g

STATUS Least Concern

This dark hummingbird has a metallic green crown, nape, and back. The male has a sooty-gray underside, a glittering green forehead, and a short, spiky crest , the color of which varies with the subspecies: Subsp. *cristatus* has a violet-tipped crest; subsp. *exilis* has an entirely green crest; subsp. *ornatus* has a blue-tipped crest; and subsp. *emigrans* has a violet-blue-tipped crest. The crestless female has white tail corners and lighter gray underparts. This species feeds at all heights in the forest strata and males hold feeding territories. In addition to nectar, it takes many insects, especially in the dry season. Males erect their crests during courtship, which includes a fast wing-clapping flight. The nest is a cup that is built low among vines or in a shrub; incubation and fledging periods are 17 to 19 days and 19 to 21 days, respectively. This is an abundant bird in much of its range.

Plovercrest

DISTRIBUTION Subsp. *lalandi*
occurs in east Brazil; subsp.
loddigesii occurs from eastern
Paraguay and northeast
Argentina to south Brazil

HABITAT Forest, scrub,
vegetated riversides; mainly
0–2,950 ft (0–900 m)

SIZE Length: 3⅜–3½ in
(8.5–9 cm). Weight: 2–3.5 g

STATUS Least Concern

Both sexes of this medium-sized, short-billed
hummingbird have a fine, pointed crest, but
that of the male is much longer, recalling the
crests of certain plover species (lapwings, genus
Vanellus). The male's plumage is otherwise green
above and shining blue on the throat, breast,
and upper belly, and white on the lower belly.
The female has entirely pale underparts. The
two subspecies have quite different plumages
in the male—subsp. *lalandi* is as described
above, while subsp. *loddigesii* has a blue rather
than green crown and crest, and the blue on
the throat and breast covers a smaller area and is
bordered with white. This bird feeds at all levels
in the forest but nests low down. Males are
territorial, and gather at leks to attract females,
giving a high, thin song of ticking notes. The
species is fairly common and accepts modified
habitats, though its current population
trend is unclear.

MALE

MALE

Sombre Hummingbird

This hummingbird was formerly classified with the sabrewings in the genus *Campylopterus*, and resembles them in general shape, having a full, square-tipped tail, long, wide wings, and a relatively long, slightly downcurved bill. The plumage in both sexes is rather grayish above, with a dark tail and wings and a slight green gloss on the back, and buffish below. This hummingbird is sometimes territorial around flowering bromeliads and other low-growing plants. Males sing from perches and flick and fan their tails when displaying at lekking sites. Studies of Sombre Hummingbird songs have helped scientists understand hummingbird vocalizations in relation to social behavior and brain structure. The female's nest is a soft, saddle-shaped construction, in which the bird incubates two eggs for 15 to 16 days. The chicks fledge at 28 days. The species is common in much of its range, and is regularly encountered in protected areas.

DISTRIBUTION An extensive area of eastern Brazil

HABITAT Forest edges, plantations; 0–2,300 ft (0–700 m)

SIZE Length: 4¾ in (12 cm). Weight: 9 g

STATUS Least Concern

Scaly-breasted Hummingbird

This is a glossy green hummingbird with a touch of white at the tips of the tail sides, and a buff underside with heavy green scaling. The rather sturdy bill is dark, with a pink base to the lower mandible. There are several subspecies, each of which shows subtle distinctions: Subsp. *maculicauda* has the least white on the tail sides; subsp. *roberti* has an entirely black bill; subsp. *furvescens* has a paler belly and darker throat than other forms; subsp. *saturatior* is the largest and darkest form; and subsp. *berlepschi* has the most white on the tail. This is an aggressive species that dominates most others at flowering trees and other rich feeding areas. It often perches to feed, and takes many insects. The nest is a sturdy cup incorporating various soft and coarse plant materials, built saddle-style across a branch. The species is locally common in some areas.

DISTRIBUTION Subsp. *cuvierii* occurs in east and central Panama; subsp. *roberti* ranges from southeast Mexico through north Guatemala and Belize to northeast Costa Rica; subsp. *maculicauda* occurs on the Pacific slope of Costa Rica; subsp. *furvescens* is found on the Pacific slope of western Panama; subsp. *saturatior* is found on Coiba Island off Panama; subsp. *berlepschi* occurs on the coast of northern Colombia

HABITAT Dry forest and forest edges, clearings, lighter woodland, secondary growth, scrub, mangroves; 0–3,950 ft (0–1,200 m)

SIZE Length: 4½–5⅛ in (11.5–13 cm). Weight: 8–10 g

STATUS Least Concern

MALE

Blossomcrown

This is an unusual and striking small bird, named for the male's crown, which grades from creamy yellow at the front through rose-pink to dark red at the back. The male is otherwise light green above and pinky gray below, with dark wings and an orange rump and tail, while the female has a darker crown and paler underside. There are two subspecies, geographically separated: subsp. *floriceps* and subsp. *berlepschi*, which has a white tail tip. The Blossomcrown feeds at low levels within the forest and is usually seen alone, but males will assemble in small groups (five or so birds) to attract females, perching close together and calling constantly. This species' status of Vulnerable reflects both its very restricted range and the intensive habitat destruction that is ongoing there, including in areas that are nominally protected.

DISTRIBUTION Subsp. *floriceps* occurs in the Santa Marta mountains in northeast Colombia; subsp. *berlepschi* occurs in the Magdalena River valley in central Colombia

HABITAT Humid primary forest and well-established secondary forest; 1,950–7,550 ft (600–2,300 m)

SIZE Length: 3⅜ in (8.5 cm). Weight: unknown

STATUS Vulnerable

MALE

Wedge-tailed Sabrewing

The Wedge-tailed Sabrewing closely resembles the Long-tailed Sabrewing but it has a shorter tail. The bill is sturdy and the outer edge of the folded wing is markedly curved (especially in males). The upperside is glossy green, the crown is blue, and the underside is dusky whitish gray. Subsp. *curvipennis* has a clearly downcurved bill, while that of subsp. *pampa* is shorter and almost straight; *pampa* also has a slightly more bluish coloration on the lower back and tail. This species mainly feeds at low and mid-levels, visiting a range of flowers and also hunting insects, mostly in flight. It is territorial. Males sing from perches, either alone or in small groups, when seeking a mate. The song combines gurgling warbles with cricket-like chirping. The species is fairly common over much of its range and is willing to accept modified habitats, visiting larger gardens, though not usually agricultural areas.

DISTRIBUTION Subsp. *curvipennis* is found in southeast Mexico; subsp. *pampa* ranges from Yucatán through Belize and north Guatemala

HABITAT Humid forest edges, secondary growth, open woodland, gardens; 0–4,600 ft (0–1,400 m)

SIZE Length: 4½–5⅜ in (11.5–13.5 cm). Weight: 4.5–7.5 g

STATUS Least Concern

MALE

Campylopterus largipennis

MALE

Gray-breasted Sabrewing

The male Gray-breasted Sabrewing is shining green above with white spots behind each eye, and dull gray underneath. It has a medium-length, slightly downcurved bill, with a black upper mandible and pink lower mandible that turns black towards the tip. The tail feathers are dark, with the outer tail feathers tipped white. The female is very similar. In addition to the nominate there are three subspecies, each with slightly different rectrices: In subsp. *aequatorialis*, the white tips are only on the last part of the rectrices; subsp. *obscurus* has gray tips to the rectrices, and it is slightly smaller than the nominate; and subsp. *diamantinensis* has larger and green rectrices. The species typically feeds at low to medium levels on the nectar of many different plants, such as Ericaceae and *Heliconia*; it also hawks for small insects. It builds a cup-like nest from plant material, including moss and lichen, which it hangs over a branch. The current global population is unknown, but it appears to be declining and it is uncommon throughout its range.

DISTRIBUTION Subsp. *largipennis* occurs in eastern Venezuela, the Guianas, and northwestern Brazil; subsp. *aequatorialis* ranges from eastern Colombia to eastern Ecuador, eastern Peru, northern Bolivia, and northwestern Brazil; subsp. *obscurus* occurs in northeastern Brazil; subsp. *diamantinensis* occurs in southeastern Brazil

HABITAT Forests, forest edges, plantations; 350–4,250 ft (100–1,300 m)

SIZE Length: 5⅛–5⅞ in (13–15 cm). Weight: 7–10.5 g

STATUS Least Concern

Rufous Sabrewing

The male Rufous Sabrewing is shining green above, with a green head, pale orange face and cheeks, and pale orange underparts. It has a medium-length, slightly downcurved black bill, and white spots behind each eye. The tail feathers are rounded; the central feathers are dark bronze and the outer tail feathers orange. The female is very similar to the male, but a little smaller. The species typically feeds at low levels in the understory on the nectar of many different plants, including *Plantago*, *Erythrina*, and *Salvia*, and it also hawks for small insects. It builds a cup-like nest from plant material such as moss and lichen on an open branch. The current global population is estimated to be fewer than 50,000 individuals, a number that appears stable, and the species is considered locally common.

DISTRIBUTION Mexico, southern Guatemala, El Salvador

HABITAT Rainforest, forest edges, plantations; 2,600–6,550 ft (800–2,000 m)

SIZE Length: 5⅛–5⅞ in (13–15 cm). Weight: 7–10.5 g

STATUS Least Concern

MALE

Violet Sabrewing

This large hummingbird, also known as De Lattre's Sabrewing, is a bright, shining purple-blue above and below in the male, with a green rump. It has a medium-length, downcurved black bill, and small white spots behind each eye. The central tail feathers are dark blue-black, and the outer tail feathers have broad white tips. The female is similar, but has green upperparts with a purple throat, gray underparts, and a more downcurved bill. The back of subsp. *mellitus* is more purple than purple-blue. The Violet Sabrewing typically feeds at low levels in the understory on the nectar of many different plants, including *Costus, Justicia* and other Acanthaceae, *Drymonia, Heliconia, Psychotria,* Bromeliaceae, and *Musa* (banana); it also catches small insects. It constructs a sturdy cup-like nest from plant fibers and moss, which it builds on a low branch overhanging a stream or ravine. The current global population is estimated at between 50,000 and 500,000 individuals, and is considered stable.

FEMALE

MALE

DISTRIBUTION Subsp. *hemileucurus* occurs in southern Mexico to south-central Nicaragua; subsp. *mellitus* occurs in Costa Rica and western Panama

HABITAT Mountainous areas, forests, forest edges, plantations, stream banks, gardens; 3,300–8,200 ft (1,000–2,500 m)

SIZE Length: 5⅛–5⅞ in (13–15 cm). Weight: 9–12 g

STATUS Least Concern

Lazuline Sabrewing

The male of this large hummingbird has a shining green back and head, with a shining, dark blue-purple throat and breast, turning to green on the flanks. It has a medium-length, slightly downcurved black bill, and small white spots behind each eye. The rounded tail feathers are reddish brown with dark green tips. The female is similar, but is gray-green underneath and has more green on the central tail feathers. This species typically feeds at low and mid-levels on the nectar of many different plants, including *Heliconia*, *Hibiscus*, and Rubiaceae, and catches insects by hawking and gleaning from foliage. It constructs a sturdy cup-like nest from plant fibers and moss, in branches overhanging water or the ground. The current global population has not been measured but is thought to be relatively stable, although the species is considered to be uncommon in Venezuela and Colombia, rare in Ecuador, and patchily distributed.

MALE

DISTRIBUTION Ranges from north-central Venezuela to northeastern Ecuador

HABITAT Mountainous areas, forests, forest edges, plantation edges, sub-paramo, gardens; 1,650–9,850 ft (500–3,000 m)

SIZE Length: 4⅜–5½ in (11–14 cm). Weight: 6–9 g

STATUS Least Concern

MALE

Campylopterus villaviscensio

Napo Sabrewing

This is a fairly large, dark hummingbird with a long, relatively heavy, and almost straight bill and mainly glossy green plumage. The male has a dark violet-blue throat patch and both sexes have blue tail sides; the female's underside is dull gray rather than green. This hummingbird feeds mainly at low and mid-levels in the forest strata, on *Heliconia* and other forest flowers. It also hawks for insects, flying out and back to a favorite perch in the manner of a flycatcher. It is territorial, like other sabrewings. This species is very little known. It is considered to be fairly common over most of its range, but its habitat is predicted to undergo severe deforestation (a 23.7 to 25.7 percent loss of forest from its range has been predicted for the period 2006–2018), hence its designation as Near Threatened. More study of its ecology and distribution is needed for effective conservation.

DISTRIBUTION South Colombia, east Ecuador, northeast Peru

HABITAT Humid montane forest; 3,300–5,900 ft (1,000–1,800 m)

SIZE Length: 5⅛–5⅜ in (13–13.5 cm). Weight: 5–9.5 g

STATUS Near Threatened

Swallow-tailed Hummingbird

A large, sturdy, and very dark-looking hummingbird, this species has a long bill and long, deeply forked tail. The sexes are similar, although the female is smaller and a little duller-toned. The head, throat, and tail are deep violet-blue, while the back and belly are shining green. Five subspecies exist: subsp. *macroura*, described above; subsp. *simoni*, which has a bluer back and belly; subsp. *cyanoviridis*, which is brighter; subsp. *hirundo*, which is drabber and shorter-tailed; and subsp. *boliviana*, which has a green head. The species is territorial, defending flower clusters at any height in the forest strata. The female usually builds a cup nest up to ten feet (3 m) above ground in a shrub or small tree, incubating the eggs for 15 to 16 days, and then tending the young for 22 to 24 days. After fledging, the young and their mother return to the nest overnight for several days. The species is common and widespread.

MALE

DISTRIBUTION Subsp. *macroura* occurs from the Guianas and much of Brazil to Paraguay; subsp. *simoni* occurs in northeast Brazil; subsp. *cyanoviridis* occurs in southeast Brazil; subsp. *hirundo* occurs in east Peru; subsp. *boliviana* occurs in northwest Bolivia

HABITAT Open countryside, savanna, gardens, plantations, forest edges; 0–4,900 ft (0–1,500 m)

SIZE Length: 5⅞–6¾ in (15–17 cm). Weight: 6–9 g

STATUS Least Concern

Stripe-tailed Hummingbird

This species is named for the white edges to its otherwise black outer tail feathers. It is a medium-sized hummingbird with a fine, slightly downcurved bill, and has a predominantly green plumage with a reddish wing patch and, in the female, gray-white underparts. Aside from the nominate, there are two subspecies, the more olive-green subsp. *nelsoni* and the whiter-tailed subsp. *egregia*. Males of the species are notably territorial and aggressive, usually defending clusters of epiphyte flowers in the canopy. Females are more likely to trapline and to visit lower strata in the forest. The nest is built quite low down in thick cover and is sometimes decorated with bright red lichens. Post-breeding, there is marked dispersal to lower altitudes, particularly in Costa Rica and Panama. The population trend is not known, but subsp. *nelsoni*, the most northerly subspecies, is likely to be under particular pressure from deforestation.

DISTRIBUTION Subsp. *eximia* occurs from east Mexico south to central Nicaragua; subsp. *nelsoni* occurs in east and southeast Mexico; subsp. *egregia* occurs in Costa Rica and west Panama

HABITAT Humid upland forest and forest borders, secondary growth; 3,300–8,200 ft (1,000–2,500 m)

SIZE Length: 3½–4⅛ in (9–10.5 cm). Weight: 4–4.5 g

STATUS Least Concern

MALE

Black-bellied Hummingbird

Rather smaller than the other *Eupherusa* species
and sometimes classified in its own genus
(*Callipharus*), this is a very dark hummingbird.
The male's head, underparts, and primaries are
velvety black, and the upperside is dark green
with a small chestnut wing patch. The female
has a mid-gray underside, and both sexes have
prominent white tail sides. The territorial males
generally feed at much higher levels than the
traplining females, especially within forests,
though there is more overlap at forest edges
and other more marginal habitats. The nest is
a small, discrete cup built under a leaf or other
sheltering overhang, and in it the female
incubates two eggs for 16 days. After breeding,
all birds move downslope. Though this bird's
range is rather small and its population trend
unclear, it seems quite common and is present
in several large protected areas.

DISTRIBUTION Caribbean
slopes of central Costa Rica
to west Panama

HABITAT Wet upland forest
and forest edges, plantations,
secondary growth;
1,300–6,550 ft (400–2,000 m)

SIZE Length: 3–3⅜ in
(7.5–8.5 cm). Weight: 3.5–4 g

STATUS Least Concern

MALE

White-tailed Emerald

A small, dainty hummingbird, this species has dark wings, a white belly and tail sides, and, in the female, a white breast and throat. Otherwise the plumage is bright, shining green. The sexes have different feeding habits, the males tending to forage from epiphyte flowers in the canopy while the females visit flowering shrubs in the understory. Though usually seen alone, males will assemble in small groups of up to five at lekking sites at the start of the breeding season in June. At this time birds are found mainly at altitudes of 3,300 to 5,600 ft (1,000 to 1,700 m), but during the rest of the year they are more widespread at both higher and lower altitudes. The breeding biology is not well described, and the bird's population status is similarly mysterious, although it is described as quite common in some protected areas in Costa Rica.

DISTRIBUTION South Costa Rica to central Panama, mainly on the Pacific slope

HABITAT Wet forest, plantations, shady gardens, mature secondary growth; 2,450–6,550 ft (750–2,000 m)

SIZE Length: 3–3⅛ in (7.5–8 cm). Weight: 3–3.5 g

STATUS Least Concern

FEMALE

Coppery-headed Emerald

This pretty, small hummingbird is one of the
very few bird species that are endemic to Costa
Rica. It is mostly glossy green, with dark wings,
white tail sides, and a strong coppery tint to the
crown and rump; the female also has a white
central throat, breast, and belly. The population
found in the Cordillera de Guanacaste is
distinctive (the male has a violet breast spot)
and may warrant designation as a subspecies.
The bird feeds from smaller flowers on trees
and shrubs, and is an enthusiastic flycatcher,
also gleaning insects from foliage. The males
form small leks (of up to six birds) to attract
mates, and the female builds a nest within
understory vegetation. After breeding, birds
tend to move to lower altitudes. The species has
a restricted range but it is common throughout,
and can be seen at hummingbird feeders in the
Monteverde Cloud Forest Reserve and other
protected areas.

FEMALE

MALE

DISTRIBUTION The Caribbean slope of Costa Rica, from the Cordillera de Tilarán to the Reventazón River valley

HABITAT Wooded habitats, including primary wet forest, established secondary growth, pastures with trees, and plantations; 1,650–4,900 ft (500–1,500 m)

SIZE Length: 3 in (7.5 cm). Weight: 3–3.5 g

STATUS Least Concern

Snowcap

This tiny hummingbird has a small, straight bill and rather short tail. The female is green above and white below, but the male is quite different and is one of the most distinctive and beautiful of all hummingbirds. Its plumage is deep, vivid reddish purple above, with a black underside and wings, and a contrasting pure white cap (subadult males show a patchy red and green pattern overall). The two subspecies are subsp. *albocoronata* and subsp. *parvirostris*; the male of the latter has a red rather than black underside. The Snowcap feeds from various small flowers, usually traplining, although males (which typically feed higher in the forest strata) sometimes attempt to hold a territory. Males sing at communal leks (up to six birds), and after mating the female constructs a tiny cup nest from tree-fern scales. Though declining as a result of deforestation in its habitat, the species is still fairly common.

FEMALE

MALE

DISTRIBUTION Subsp. *albocoronata* occurs in west-central Panama (both slopes); subsp. *parvirostris* occurs on the Caribbean slopes of southern Honduras, Nicaragua, Costa Rica, and far west Panama

HABITAT Wet forest and forest edges, secondary growth; 0–5,250 ft (0–1,600 m)

SIZE Length: 2⅜–2½ in (6–6.5 cm). Weight: 2.5 g

STATUS Least Concern

White-vented Plumeleteer

A fairly large, long-billed hummingbird, this species has a shining green plumage with a dark blue tail and conspicuous white undertail. In addition, the female has a gray-white throat, breast, and belly. Aside from the nominate subsp. *buffonii*, described above, there are three subspecies: subsp. *micans*, which is larger and has a bluer tail; subsp. *aeneicauda*, which is more yellow-green; and the colorful subsp. *caeruleogaster*, in which the male has blue rather than green underparts. The White-vented Plumeleteer is an aggressive species, capable of driving off most other hummingbirds, though it is not always territorial and sometimes traplines. It visits clumps of flowers of a wide range of species and may steal nectar from flowers with long corollas by piercing them at their base. It is also an adept flycatcher and takes insects from spiders' webs. The species is quite common and is increasing in most of its range, although subsp. *caeruleogaster* is thought to be in decline.

MALE

JUVENILE

DISTRIBUTION Subsp. *buffonii* occurs in central to northeast Colombia and northwest Venezuela; subsp. *micans* occurs in central and eastern Panama and west Colombia; subsp. *aeneicauda* occurs from north Colombia to west and north-central Venezuela; subsp. *caeruleogaster* occurs in north and central Colombia

HABITAT Dry and damp forest and forest edges, secondary growth, dense scrub; 0–6,550 ft (0–2,000 m)

SIZE Length: 4⅛–4¾ in (10.5–12 cm). Weight: 6–7.5 g

STATUS Least Concern

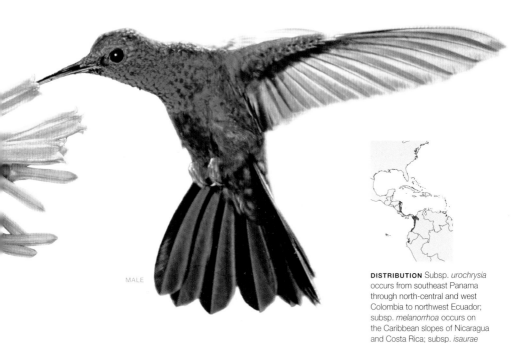

MALE

Chalybura urochrysia

Bronze-tailed Plumeleteer

This species is very similar to the White-vented Plumeleteer but the male has a dark bronzy-brown tail and both sexes have bright red feet. There are four subspecies: subsp. *urochrysia*; subsp. *melanorrhoa*, which has a dark rather than white vent; subsp. *isaurae*, which has blue on the throat and breast; and subsp. *intermedia*, which has a bluish tail and breast. Like the White-vented, this is an aggressive bird, dominant at hummingbird gatherings, and is sometimes territorial, particularly around stands of *Heliconia* flowers. Females visit males'

territories at the start of the breeding season, to access good nectar stores as well as to mate. The nest is usually built low down in the forest overhanging a stream, and is a deep, thickly lined cup with outer decorations of moss and lichen. The species is quite common locally but has declined in numbers overall. It will, however, use modified habitats such as gardens.

DISTRIBUTION Subsp. *urochrysia* occurs from southeast Panama through north-central and west Colombia to northwest Ecuador; subsp. *melanorrhoa* occurs on the Caribbean slopes of Nicaragua and Costa Rica; subsp. *isaurae* occurs on Caribbean slopes from Panama to northwest Colombia; subsp. *intermedia* occurs in southwest Ecuador

HABITAT Damp forest, forest edges, secondary growth, other well-vegetated, shady habitats; 0–2,950 ft (0–900 m)

SIZE Length: 4⅛–4¾ in (10.5–12 cm). Weight: 6–7 g

STATUS Least Concern

Mexican Woodnymph

The male Mexican Woodnymph is shining green above, with a blue-green cap on the back of its head and a bright purple-blue forehead. Its throat is a bright green, turning to darker green underneath. It has a medium-length, straight black bill. The dark blue-black tail feathers are slightly forked. The female is very similar, but has small white spots behind each eye, gray underparts, and white tips on the outer tail feathers. The species is thought to feed on the nectar of a range of different plants, including Rubiaceae, Zingiberaceae, Bromeliaceae, and Ericaceae, and also hawks for small insects. However, very little is known about the Mexican Woodnymph's ecology, including its food sources, habitat requirements, and breeding behavior. The current global population is thought to be between 10,000 and 20,000 individuals, and is in decline as a result of habitat loss.

DISTRIBUTION Mexico

HABITAT Forests, canyons, plantations, stream banks; 350–4,250 ft (100–1,300 m)

SIZE Length: 3⅛–3⅞ in (8–10 cm). Weight: 3–4.5 g

STATUS Vulnerable

MALE

Crowned Woodnymph

Also known as the Blue-, Purple-, or Violet-crowned Woodnymph, the Crowned Woodnymph now includes the Green-crowned Woodnymph (subspp. *fannyae*, *subtropicalis*, *verticeps*, and *hypochlora*), which was previously considered a separate species. Subsp. *hypochlora* has also sometimes been considered separate— the Emerald-bellied Woodnymph. Males are mainly glittering green and violet with a deeply forked blue-black tail. The subspecies differ in the color of the crown—brilliant violet in subspp. *townsendi*, *venusta*, *colombica*, and *rostrifera*, glittering green in *fannyae*, *subtropicalis*, *verticeps*, and *hypochlora*. The male *hypochlora* has a green rather than violet belly. Females have gray underparts, darker on the belly. Crowned Woodnymphs feed on the nectar of a wide range of flowers, including *Heliconia*, Ericaceae, Bromeliaceae, and Acanthaceae, and also hawk for small insects. Males are aggressive and often territorial. The cup-like nest is built from plant material on a branch, beneath a leaf. The current global population is unknown, and although the species is thought to be declining in number, it is regarded as common.

FEMALE

MALE

DISTRIBUTION Subsp. *colombica* ranges from northern Colombia to northwestern Venezuela; subsp. *townsendi* ranges from northeastern Guatemala to eastern Honduras; subsp. *venusta* ranges from eastern Nicaragua to central Panama; subsp. *rostrifera* occurs in western Venezuela; subsp. *fannyae* ranges from eastern Panama to western Colombia; subsp. *subtropicalis* occurs in west-central Colombia; subsp. *verticeps* ranges from southwestern Colombia to western Ecuador; subsp. *hypochlora* ranges from southwestern Ecuador to northwestern Peru

HABITAT Forests, forest edges, plantations, gardens; 0–6,550 ft (0–2,000 m)

SIZE Length: 3⅛–4⅜ in (8–11 cm). Weight: 3–4.5 g

STATUS Least Concern

Fork-tailed Woodnymph

Also known as the Common Woodnymph, this species is separated into 13 subspecies, the males of which share a shining green head and back, a brilliant green throat, and a shining purple-blue belly and patch at the top of the back. The medium-sized black bill is slightly downcurved, and the blue-black tail is deeply forked. Females have a duller green head and gray underparts, and the outer tail feathers have white tips.

The subspecies display slight variations in coloration and size. Fork-tailed Woodnymphs feed on the nectar of a wide range of flowers, including *Heliconia*, Bromeliaceae, Ericaceae, Rubiaceae, and Acanthaceae, as well as on small insects. Males are aggressive and are often territorial. Little is known about the species' breeding behavior. The current global population is unknown, and although the species is thought to be declining in number, it is regarded as common.

MALE

DISTRIBUTION Subsp. *furcata* occurs in eastern Venezuela, the Guianas, and northeastern Brazil; subsp. *refulgens* occurs in northeastern Venezuela; subsp. *fissilis* occurs in eastern Venezuela, western Guyana, and northeastern Brazil; subsp. *orenocensis* occurs in southern Venezuela; subsp. *nigrofasciata* occurs in southeastern Colombia, northwestern Brazil, and southern Venezuela; subsp. *viridipectus* occurs in eastern Colombia, eastern Ecuador, and northeastern Peru; subspp. *jelskii* and *simoni* occur in eastern Peru and western Brazil; susbsp. *balzani* occurs in north-central Brazil; subsp. *furcatoides* occurs in eastern Brazil; subsp. *boliviana* occurs in southeastern Peru and northeastern Bolivia; subsp. *baeri* ranges from northeastern and central Brazil to southeastern Bolivia and northwestern and north-central Argentina; subsp. *eriphile* occurs in southeastern Brazil, Paraguay, and northeastern Argentina

HABITAT Forests, forest edges, semi-open areas, plantations, gardens; 0–6,550 ft (0–2,000 m)

SIZE Length: 3⅛–4⅜ in (8–11 cm). Weight: 4–6 g

STATUS Least Concern

Violet-capped Woodnymph

The male Violet-capped Woodnymph is shining, dark gold-green above and green below, with a purple-blue cap and forehead, and a bright green throat. It has a medium-length, straight black bill, and the dark blue tail is forked. The female is very similar, but is a paler, dull green above, lacks the purple-blue cap and forehead, and has pale gray underparts and a less pronounced fork in the tail. The species feeds on the nectar of a wide range of different plants, including Bromeliaceae, Orchidaceae, Passifloraceae, Fabaceae, Rubiaceae, and Euphorbiaceae, and eats small insects. It builds a small cup-like nest from soft plant fibers, moss, and lichen, positioned on a tree branch. The current global population and population trend are unknown, but the species is regarded as common.

DISTRIBUTION Southern and eastern Brazil, eastern Paraguay, northeastern Argentina

HABITAT Forests, forest edges, scrubland, parks, gardens, urban areas; 0–3,300 ft (0–1,000 m)

SIZE Length: 3⅛–4¾ in (8–12 cm). Weight: 3–5 g

STATUS Least Concern

MALE

Many-spotted Hummingbird

DISTRIBUTION The eastern Andes from east Ecuador and east Peru through central and southeast Bolivia to northwest Argentina

HABITAT Steep forest interiors and edges, including secondary growth; 1,300–3,300 ft (400–1,000 m)

SIZE Length: 4⅛–4½ in (10.5–11.5 cm). Weight: 6.5–7 g

STATUS Least Concern

This quite large but slender hummingbird has a long, fine, slightly downcurved bill. The sexes are alike, with dark wings and otherwise shining green upperparts. The underside is whitish but heavily marked with neat, distinct green spots (there is slightly less spotting in the female). The species forages at low levels within thick vegetation, visiting many flower types, including *Passiflora*, *Inga*, and *Palicourea*. Insect-hunting is conducted in more open situations, the bird flying out from a perch to catch flies, often over streams. The substantial nest, made from rather coarse materials, is fixed to a tree trunk, barely above ground level. The incubation and fledging periods are 14 to 15 days and 20 to 22 days, respectively. There is much deforestation ongoing in the species' range and it is declining, but it remains common in most areas. More surveys are needed to clarify its status.

FEMALE

White-throated Hummingbird

This striking hummingbird has a long, fine bill and bright, glittering green plumage, with dark wings, a broad white throat patch, and a white lower belly, undertail, and tail corners. There is a touch of orange gloss on the rump. Females are similar to males but less shiny. A widespread species with wide-ranging tastes in nectar, this bird feeds at low and high levels in the forest strata and flycatches over open areas. It nests quite low in a tree or shrub, building a cup nest from lichens, cobwebs, soft plant parts, and moss. Incubation takes 14 days and fledging 20 to 25 days. The species' taste for introduced plants such as *Citrus* and *Salvia* allow it to thrive in quite extensively modified habitats, and it is a common garden bird in some areas, especially in the south and east of its range.

DISTRIBUTION Eastern Bolivia and eastern Paraguay to Uruguay, northern Argentina, and southeast Brazil

HABITAT Forest edges, marshland, parks, gardens; 0–3,300 ft (0–1,000 m)

SIZE Length: 3⅞–4½ in (10–11.5 cm). Weight: 4.5–5 g

STATUS Least Concern

MALE

Buffy Hummingbird

The sexes are almost alike in this fairly small, slight hummingbird, both having rather subdued light olive-green upperparts and pale underparts with a variable pinkish-peach flush on the breast and belly. Females are typically a little duller than the males. The Buffy Hummingbird feeds at low levels from flowers such as *Agave*, various cacti, and *Hibiscus*, and is usually seen alone. It breeds in the rainy season and builds a saddle-shaped nest on a thin branch or fork about six feet (2 m) above ground, using mainly soft fibers from *Gossypium* seeds. Post-breeding, it tends to move away from the most arid areas to thicker thorn forest. It is a very common bird in its habitat and is frequently found close to human habitation, though extensive urban development along this coastline may threaten its long-term prospects.

DISTRIBUTION Coastal Colombia and Venezuela, and adjacent islands

HABITAT Dry, open, thorny scrubland and thorn forest, gardens, parks, mangroves; 0–1,800 ft (0–550 m)

SIZE Length: 3⅜–3½ in (8.5–9 cm). Weight: unknown

STATUS Least Concern

MALE

Leucippus baeri

Tumbes Hummingbird

This is a very plain brownish-gray hummingbird with a fine, slightly downcurved bill. It shows only a slight olive-green gloss on the lower back and rump, and slightly warmer golden tones on the head. The wings are darker, and the underside is off-white. The male and female look alike. This species is very little known. It feeds at low levels, particularly from flowering cacti, and also hawks for flying insects. Its breeding habits are presumed to be similar to those of the other species in its genus. Birdwatchers visiting the area where it occurs have reported that it is fairly common, but the exact limits of its distribution are not known with any certainty. Various potential threats may affect its habitat, and surveys are needed to gain a clear picture of its distribution and current population trend.

DISTRIBUTION Southwest Ecuador to northwest Peru along the coast

HABITAT Dry forest and scrub, deciduous forest; 0–4,250 ft (0–1,300 m)

SIZE Length: 3½–3⅞ in (9–10 cm). Weight: 4.5 g

STATUS Least Concern

MALE

Spot-throated Hummingbird

This drab-colored species is like a slightly larger version of the Tumbes Hummingbird, but the whitish underparts are marked with fine olive-green spots on the throat and upper breast. The sexes look alike. As is typical with *Leucippus* species, this is a low-level feeder in arid habitats, with a particular liking for cactus flowers, although it will also visit introduced species such as banana (*Musa*) and *Agave*, and it regularly hawks for flying insects. Although it sometimes ranges to high altitudes, it is usually found below 3,300 feet (1,000 m). The bird's habits are little known, though it has been described as being fairly common within its limited range. Its current population trend is also unknown, but its acceptance of modified habitats and introduced nectar sources should hold it in good stead for long-term survival.

DISTRIBUTION Western Andean slopes in north and central Peru

HABITAT Dry forest, scrubland, plantations; 1,150–9,850 ft (350–3,000 m)

SIZE Length: 4½–4⅞ in (11.5–12.5 cm). Weight: 7–7.5 g

STATUS Least Concern

Olive-spotted Hummingbird

A straight-billed *Leucippus* species, this bird is similar to the Spot-throated Hummingbird but has rather duller, more yellow-olive upperparts and finer spotting on the underside, mainly restricted to the throat. It feeds at mainly small flowers at all heights in the forest strata, hovering rather than clinging, and often catches flies. Usually seen alone, it is territorial and aggressive to others of its species and similar-sized hummingbirds. It may even closely approach a human observer. The nest is a small, soft cup, built up to 33 feet (10 m) above ground on a horizontal branch. The incubation and fledging periods are 14 to 15 days and 20 days, respectively. The species' numbers are falling, a trend predicted to continue as around 15 percent of its habitat is expected to be lost to deforestation over the next 12 years.

DISTRIBUTION Far southeastern corner of Colombia, east Ecuador, northeast Peru, northwest Brazil

HABITAT Amazonian tropical forest and forest clearings; 0–500 ft (0–150 m)

SIZE Length: 4¾ in (12 cm). Weight: 6 g

STATUS Least Concern

MALE

White-bellied Hummingbird

This small, straight-billed hummingbird has shining, light green upperparts and white underparts, with green spotting on the throat sides and flanks. The more northerly subsp. *chionogaster* has pure white underparts while those of subsp. *hypoleuca* have a creamy tint. The species resembles members of *Leucippus* in general appearance and was until recently placed in that genus. A bird with diverse tastes, it visits all kinds of flowers at all heights in the forest strata, and also eats many insects. The nest is a soft, thickly lined cup, built in a large shrub or small tree. The two eggs are incubated for 14 to 15 days. The chicks are fed in the nest by just the female for 19 to 22 days, and care continues for several days after fledging. In the far south of the range there is some post-breeding dispersal. The species' population size and trend are unknown.

MALE

DISTRIBUTION Subsp. *chionogaster* is found in north and central Peru; subsp. *hypoleuca* ranges from southeast Peru through Bolivia to northwest Argentina

HABITAT A wide range of mainly dry but well-vegetated habitats, including forest edges, secondary growth, plantations, scrub, cerrado, and gardens; 1,500–9,200 ft (450–2,800 m)

SIZE Length: 3½–4¾ in (9–12 cm). Weight: 4.5–5.5 g

STATUS Least Concern

Green-and-white Hummingbird

This species is very similar to the White-bellied Hummingbird, but has more olive-toned upperparts, some greenish-gray streaking on the vent, and a wholly dark underside to the tail. Like that species it was formerly placed in the genus *Leucippus*. It is a little-known species that appears to visit a good range of flower species and catches flies on the wing. The nest is a relatively thick cup and incorporates moss, spider webs, soft plant fibers, and lichen, and is usually built on a thin branch in a small tree. The species' population trend is not known, and surveys are needed to determine any appropriate conservation measures. Although this hummingbird has a very restricted range, it is quite likely to be seen by tourists visiting Machu Picchu in Peru, where it is a common breeding species.

DISTRIBUTION Central Peru on eastern Andean slopes

HABITAT Secondary growth, glades and clearings, forest edges; 3,300–8,200 ft (1,000–2,500 m)

SIZE Length: 3⅞–4⅜ in (10–11 cm). Weight: 5.5–6 g

STATUS Least Concern

MALE

Amazilia tzacatl

Rufous-tailed Hummingbird

A common and widespread species of the northern tropics, the noisy and aggressive Rufous-tailed Hummingbird comprises four subspecies that can be found sparring over feeding rights in a wide range of habitats and elevations from eastern Mexico to western Ecuador. Birds are not choosy about nectar sources and take advantage of flowers of many sizes and shapes, including insect-pollinated species such as Coral Vine (*Antigonon leptopus*) and lantanas (*Lantana*), and cultivated species such as bananas (*Musa*) and ornamental *Hibiscus*. They also accept sugar-water feeders and often dominate feeding stations. Females are slightly drabber than males, with darker bills and narrow pale fringes on the throat feathers. They nest in the shrubby understory, building a thin-walled cup of fine plant fibers and spider silk camouflaged with mosses, lichens, and other plant fragments. The species name means "grass" in Nahuatl, the language of the Aztec Empire.

DISTRIBUTION Subsp. *tzacatl* ranges from central-eastern Mexico to central Panama; subsp. *fuscicaudata* occurs in northern and western Colombia and western Venezuela; subsp. *handleyi* is restricted to Escudo de Veraguas Island off the coast of northwestern Panama; subsp. *brehmi* is found in southwest Colombia; subsp. *jucunda* ranges from northwest Colombia to southwest Ecuador

HABITAT Tropical evergreen forest, forest edges, and clearings, secondary growth, plantations, gardens; 0–8,200 ft (0–2,500 m)

SIZE Length: 3½–4⅜ in (9–11 cm). Weight: 5–7 g

STATUS Least Concern

Chestnut-bellied Hummingbird

This hummingbird has green upperparts that shade through olive to chestnut orange on the tail. The throat and upper breast are bright green, and the belly chestnut. The flight feathers are blackish. An enigmatic bird, this species has a very restricted range and is declining, but appears to be quite versatile in its habits. Analysis of pollen taken from the feathers of feeding birds have shown that at least 15 different plant species are regularly visited, and the hummingbird readily travels to modified habitats to feed. However, it may be less willing to breed in such habitats, and is often absent from apparently suitable habitat patches for no obvious reason. Its population has been extensively surveyed and is estimated at 1,200 individuals. It is known to occur in 14 sites, a few of which fall within protected areas. Conservation work is underway, with much involvement from local communities.

MALE

DISTRIBUTION Northern-central Colombia, on the western slopes of the Andes

HABITAT Forest edges on submontane slopes, scrubby ravines, sometimes plantations and other cultivated land; 1,650–6,550 ft (500–2,000 m)

SIZE Length: 3½ in (9 cm). Weight: unknown

STATUS Endangered

Buff-bellied Hummingbird

The aggressive and adaptable Buff-bellied Hummingbird is a year-round resident of native woodlands and thickets and developed areas from the Yucatán Peninsula northward along the Gulf of Mexico to southeastern Texas. The sexes are similar, though females have darker bills and delicate ivory fringes on the green feathers of the chin and throat. The belly is pale buff in the northern subsp. *chalconota*, darker in subsp. *cerviniventris*, and rich cinnamon rufous in subsp. *yucatanensis.* Though the birds feed on a variety of nectar sources and have been observed stealing sap from sapsucker wells in winter, much of their foraging time is spent in pursuit of insects and other invertebrates. Feeding stations and winter-blooming ornamental flowers such as Shrimp Plant (*Justicia fulvicoma*) and cigar plants (*Cuphea*) have allowed individuals from the northern subspecies to overwinter successfully along the Gulf Coast of the southeastern United States well north of its breeding range.

DISTRIBUTION Subsp. *yucatanensis* occurs in northern Belize, Mexico's Yucatán Peninsula, and northern Guatemala; subsp. *cerviniventris* occurs in southern Mexico; subsp. *chalconota* is found in the southern United States (breeds in southern Texas, winters as far east as Florida) and northeastern Mexico

HABITAT Oak, mesquite, and acacia woodlands, tropical deciduous forest, thorn scrub, palm thickets, secondary growth, parks, gardens; 0–3,950 ft (0–1,200 m)

SIZE Length: 3¾–4⅜ in (9.5–11 cm). Weight: 3.5–4.5 g

STATUS Least Concern

MALE

Cinnamon Hummingbird

The bright cinnamon-buff underparts that give the species its name distinguish Cinnamon Hummingbirds from all other members of the widespread genus *Amazilia*. The plumage is similar in both sexes and all ages, though the bills of females and young males have more extensive black at the tip. Subsp. *diluta* of northwestern Mexico has paler underparts than those of the more southerly and eastern subspp. *rutila* and *corallirostris*. The fourth subspecies, *graysoni*, a resident of the Marías Islands off the Pacific coast of Mexico, is slightly larger and duller than birds of the nearby mainland and at times has been considered a separate species. Northern populations are partially migratory, and strays have reached the southwestern United States. Like many other *Amazilia* species, they are territorial and noisily defend abundant nectar sources, including a wide variety of trees, shrubs, vines, epiphytes, cacti, thistles, and tropical mistletoes (*Psittacanthus*).

DISTRIBUTION Subsp. *diluta* occurs in northwestern Mexico; subsp. *rutila* occurs in western and southwestern Mexico; subsp. *corallirostris* occurs in southern and southeastern Mexico; subsp. *graysoni* is restricted to the Marias Islands off the western coast of Mexico

HABITAT Tropical deciduous and semideciduous forest, woodland, and forest edges, thorn forest, secondary growth, savanna, coastal scrub, coconut groves, coffee plantations, gardens; 0–5,400 ft (0–1,650 m)

SIZE Length: 3⅞–4¾ in (10–12 cm). Weight: 4.5–5 g

STATUS Least Concern

MALE

Amazilia Hummingbird

The Amazilia Hummingbird has a gold-green head, a reddish back and underparts, and a shining green throat. The medium-length, slightly downcurved bill is reddish with a black tip, and the slightly forked tail is reddish with a dark band at the tips. The female is similar to the male, but is paler underneath and has white markings on the throat. Six subspecies have been identified: Subsp. *dumerilii* is smaller than the nominate and has a white throat patch; subsp. *leucophoea* is also smaller with a white throat patch, but its upperparts are bronzy green instead of reddish; subsp. *caeruleigularis* has a purple throat patch; subsp. *alticola* has less rufous on the chest; and subsp. *azuay* is similar to *alticola* but with whiter belly. The species feeds on the nectar of a range of flowering plants, including *Erythrina* and *Psittacanthus*, as well as on small insects and spiders. It builds a cup-like nest from soft plant material on an overhanging branch or in thick scrub. The total population and population trend are unknown, but the species appears to be stable and is regarded as common.

MALE

DISTRIBUTION Subsp. *amazilia* occurs in western Peru; subsp. *dumerilii* ranges from southeastern and western Ecuador to northern Peru; subsp. *leucophoea* occurs in northwestern Peru; subsp. *caeruleigularis* occurs in southwestern Peru; subsp. *alticola* is found in southern Ecuador; subsp. *azuay* occurs in south-central Ecuador

HABITAT Semiarid to arid open areas, coasts, scrubland, thorn forest, desert, steppe; 0–9,850 ft (0–3,000 m)

SIZE Length: 3⅛–4⅜ in (8–11 cm). Weight: 4–6 g

STATUS Least Concern

Plain-bellied Emerald

One of the *Amazilia* species that lacks strong orange or brown tones, this species is white below and shining yellow-green above with a dark tail and wings. As with other members of the genus, the sexes are similar. Subsp. *bahiae* has a longer bill and longer wings than the nominate. The Plain-bellied Emerald is a bird of coastal areas, and mainly feeds quite low in the forest strata, taking nectar from small herbaceous plants as well as shrubs and trees. It is territorial and aggressive, giving a high, thin call as it chases away rivals. When not taking nectar it flycatches, and may rest on a prominent high perch. The nest is a soft cup made from woolly plant seeds and fibers, usually situated rather low in a tree and often overhanging water. This is a rather common bird but the exact distribution of its two subspecies is unclear.

DISTRIBUTION Subsp. *leucogaster* is found in eastern Venezuela, through the Guianas into northeast Brazil; subsp. *bahiae* continues along the coast of eastern Brazil as far as Bahia

HABITAT Mangroves, grassland, forest edges, plantations; 0–800 ft (0–250 m)

SIZE Length: 3½–3⅞ in (9–10 cm). Weight: 4.5 g

STATUS Least Concern

MALE

Versicolored Emerald

This hummingbird has a very large geographic range and shows significant subspecific variation. The nominate subsp. *versicolor* is shining green with a pale belly patch; subsp. *millerii* has a shining, light blue crown and white throat patch; subsp. *hollandi* is the most striking form, with a white throat and mostly blue head; subsp. *nitidifrons* has a turquoise to golden crown; subsp. *rondoniae* has a blue to green crown; and subsp. *kubtcheki* has a shorter bill and shining green throat. The bird feeds at all levels on a huge range of flower types, and like other *Amazilia* species is normally aggressive and territorial. It builds the typical *Amazilia* cup nest on a horizontal branch, using bromeliad seeds and other soft materials in its construction. Incubation and fledging periods are 14 and 20 to 26 days, respectively. The status of the various subspecies is not clearly known, but the bird appears to be scarcer and more patchily distributed in the north of its range.

MALE

DISTRIBUTION Subsp. *versicolor* occurs in southeast Brazil; subsp. *millerii* is the most northerly form, found in tropical Colombia, central Venezuela, and north Brazil; subsp. *hollandi* occurs in southeast Venezuela and western Guyana; subsp. *nitidifrons* occurs in northeast Brazil; subsp. *rondoninae* occurs in west-central Brazil; subsp. *kubtcheki* ranges from northeast Bolivia through east Paraguay to extreme northeast Argentina and southwest Brazil

HABITAT Rainforest in the north of the range, more open habitats such as scrub, plantations, and gardens farther south; 150–5,600 ft (50–1,700 m)

SIZE Length: 3⅛–3⅞ in (8–10 cm). Weight: 3.5–4 g

STATUS Least Concern

White-chested Emerald

This shining green hummingbird has a coppery flush on the rump and tail, and white underparts with green spotting on the sides of the throat, breast, and flanks. Subsp. *chionopectus* is larger than the other subspecies, and the very rare (possibly extinct) subsp. *orienticola* is darker and more bronze-toned than the nominate. This species is best known from Trinidad, where it is one of the commonest hummingbirds and uses more open habitats than on the mainland. It feeds mainly from flowers on large trees, and takes insects from foliage as well as in flight. It builds a typical cup-shaped nest on a horizontal branch. Subsp. *orienticola* has been observed very infrequently and not for several decades—surveys are needed to establish whether it still survives and to ascertain the status of the other mainland form, subsp. *brevirostris*.

DISTRIBUTION Subsp. *brevirostris* occurs in eastern Venezuela, Guyana, Suriname, and the far north of central Brazil; subsp. *chionopectus* is found on Trinidad; subsp. *orienticola* is recorded from the coast of French Guiana

HABITAT Forest and forest edges, secondary growth, scrubland, scrubby grassland, plantations; 0–1,650 ft (0–500m)

SIZE Length: 3½–3⅞ in (9–10 cm). Weight: 4.5 g

STATUS Least Concern

MALE

Amazilia franciae

Andean Emerald

The Andean Emerald has a shining purple-blue cap and forehead; bright green cheeks and neck; shining green upperparts, turning bronze towards the rump; and white underparts. The medium-length, straight bill has a black upper mandible and black-tipped pink lower mandible. The tail feathers are bronze. The female is similar to the male, but the cap is duller and more green. Subsp. *viridiceps* has a shorter tail and green cap, while in subsp. *cyanocollis* the purple-blue of the cap extends down the back of the head. The species feeds by traplining mostly at mid- to high levels on the nectar of a wide range of plants, including *Musa* (banana), *Canna*, *Psammisia*, and *Eugenia*; it also catches small insects. It builds a small cup-shaped nest from soft plant material. It is most common above 3,300 feet (1,000 m). The total population and population trend are unknown, but the species appears to be stable and is regarded as fairly common.

MALE

DISTRIBUTION Subsp. *franciae* occurs in northwestern and central Colombia; subsp. *viridiceps* occurs in Ecuador and southwestern Colombia; subsp. *cyanocollis* occurs in northern Peru

HABITAT Forest edges, clearings, arid areas, shrubland; 1,650–6,550 ft (500–2,000 m)

SIZE Length: 3⅛–4⅜ in (8–11 cm). Weight: 5–6 g

STATUS Least Concern

White-bellied Emerald

The White-bellied Emerald resembles the Azure-crowned Hummingbird but it is smaller and has all white underparts. It has a dark green head and shining green upperparts. The shining blue tail has a slight fork. The short, straight bill is black, with pink at the base of the lower mandible. The tail feathers are green with a dark band across their tips. The female is similar to the male, but has mottling on the sides of the throat and gray tips to the tail feathers. Subspp.

genini and *pacifica* have much longer bills than the nominate, and subsp. *pacifica* also has longer wings. The species feeds at all levels on the nectar of a wide range of plants, including *Inga*, *Miconia*, *Psychotria*, and *Heliconia*, as well as on small insects. It builds a small cup-like nest from soft plant material. The current global population appears to be declining as a result of deforestation, but the species is still regarded as very common.

DISTRIBUTION Subsp. *candida* occurs in southeastern Mexico and southern Guatemala; subsp. *genini* occurs from eastern to southern Mexico; subsp. *pacifica* ranges from southeastern Mexico to Belize, northern Guatemala, Honduras, and Nicaragua

HABITAT Forest, riverbanks, clearings, plantations; 1,000–4,900 ft (300–1,500 m)

SIZE Length: 3⅛–4⅜ in (8–11 cm). Weight: 3–4 g

STATUS Least Concern

FEMALE

Mangrove Hummingbird

The Mangrove Hummingbird is a medium-sized hummingbird, with green upperparts that grade to bronze on the tail, whitish underparts with green mottling at the breast and sides, and a shining blue-green throat. The medium-length bill is slightly downcurved, with a dark-tipped reddish lower mandible. The slightly forked tail is dark brown. The female is similar to the male, but has more white below with some blue markings on the sides of the throat. The species feeds on the nectar of a range of plants, including *Pelliciera* and *Lonchocarpus*. In the dry season, it ventures away from mangroves to adjacent habitat to feed at *Inga* and other trees such as *Tabebuia ochracea* and *T. impetiginosa*; it also hawks for mosquitoes. The small cup-like nest is built from soft plant material on top of a mangrove twig. The population is very local to Costa Rica's Pacific coast and is estimated at 2,500 to 10,000; it is thought to be declining as a result of the clearance and degradation of the species' mangrove habitat.

DISTRIBUTION Costa Rica

HABITAT Coastal areas, mangroves, sandbars; 0–350 ft (0–100 m)

SIZE Length: 3½–4⅜ in (9–11 cm). Weight: 4–5 g

STATUS Endangered

MALE

MALE

Amazilia amabilis

Blue-chested Hummingbird

An olive-green hummingbird, this species has bright, glittering green feathers on the forehead and chin, and a diffuse light blue patch on the lower throat. The upperparts shade through olive to brown on the tail. The female has a speckled whitish throat. This is typically a solitary feeder that visits small, low-growing flowers, though on occasion groups may assemble at flowering trees, which they attempt to defend from other hummingbirds. The males form loose gatherings at the start of the breeding season, each singing from a perch to attract females. The cup-shaped nest is well concealed in a shrub or small tree, and incorporates tree-fern scales and cobwebs, as well as moss and lichen.

When not breeding, the species undertakes unpredictable short-range movements in response to changes in food supplies. It is fairly common to very common in most areas.

DISTRIBUTION Northeast Nicaragua to north-central Colombia and Ecuador

HABITAT Lightly wooded areas, forest edges and glades, plantations, sometimes gardens; 0–3,300 ft (0–1,000 m)

SIZE Length: 2¾–3⅞ in (7–10 cm). Weight: 4 g

STATUS Least Concern

MALE

Charming Hummingbird

This species was until recently considered a subspecies of the Blue-chested Hummingbird and is very similar to that bird, but is slightly larger and longer-winged. Birds have a glittering green head and vivid blue throat patch, with otherwise rather dark olive to brown plumage. There is little difference between the sexes, though females are somewhat duller and have pale, speckled throats. The species also closely resembles the Blue-chested in terms of its habits, being primarily a solitary trapliner. It sometimes forms territories around flowering trees; these are unrelated to the patches occupied by males during courtship. It is inclined to move to higher altitudes after the breeding season. This bird has a localized distribution and is considered to be a restricted-range species. Its population size and trend are not yet known but it is generally regarded as fairly common.

DISTRIBUTION Southwest Costa Rica, Pacific west Panama

HABITAT Lowland and submontane forest edges and glades, plantations, secondary growth; 0–4,600 ft (0–1,400 m)

SIZE Length: 3½–4⅜ in (9–11 cm). Weight: 4 g

STATUS Least Concern

Amazilia cyanocephala

Azure-crowned Hummingbird

This is a light green *Amazilia* species with a mostly dark bill in both sexes. The underside is white. Subsp. *cyanocephala*, which may qualify for full species status, has a light turquoise crown, while that of subsp. *chlorostephana* is green, concolorous with the rest of the head. The nominate has much more generalist habitat preferences than subsp. *chlorostephana*, which occurs below 350 feet (100 m) in areas of pine savanna. Birds feed mainly at mid- to high levels in the forest on flowers such as *Inga*, and sometimes in the understory on *Heliconia*; they also hawk for insects. The species builds an open cup nest, usually quite high in a tree. Subsp. *cyanocephala* tends to breed at higher elevations and then move downslope after breeding. It is quite common throughout its range, while subsp. *chlorostephana* is very localized but seems to have benefited from forest clearance.

MALE

DISTRIBUTION Subsp. *cyanocephala* ranges from east and south Mexico to east Honduras and northeast Nicaragua; subsp. *chlorostephana* occurs along the Mosquito Coast of east Honduras and northeast Nicaragua

HABITAT A very wide range of habitats, including cloud forest, rainforest edges, pine and pine–oak woodland, gardens, plantations and disused farmland; 0–7,850 ft (0–2,400 m)

SIZE Length: 3⅞–4⅜ in (10–11 cm). Weight: 3.5–4 g

STATUS Least Concern

Berylline Hummingbird

A two-toned hummingbird, this species has a shining green head, back, and breast, and a chestnut-copper lower back, rump, and tail. There are also chestnut markings in the otherwise dark wings. There are five subspecies: In subsp. *beryllina* the lower back is green; in subsp. *viola* the back and rump have a gray tinge; in subsp. *lichtensteini* the green areas are paler; in subsp. *sumichrasti* there is a violet tinge to the rump and tail; and in subsp. *devillei* the back and rump are a richer chestnut-copper color. This hummingbird feeds at all heights in the forest strata and is often observed joining other hummingbirds at flowering trees, where it dominates most other species. It is migratory in the far north of its range (subsp. *viola*).

DISTRIBUTION Subsp. *beryllina* is found in central Mexico; subsp. *viola* occurs in north and northwest Mexico; subsp. *lichtensteini* occurs in west Chiapas, south Mexico; subsp. *sumichrasti* occurs in central and south Chiapas, south Mexico; subsp. *devillei* ranges from south Guatemala and El Salvador to central Honduras

HABITAT Tropical forest edges, dense oak and pine–oak woodland, scrub, thorn forest, plantations, gardens; 0–9,850 ft (0–3,000 m)

SIZE Length: 3⅛–3⅞ in (8–10 cm). Weight: 4–4.5 g

STATUS Least Concern

MALE

Blue-tailed Hummingbird

A pretty hummingbird with mainly shining green plumage, this species also shows a small bronze-chestnut wing patch, golden rump, and steely-blue tail. The feathers at the base of the legs are white. There are three subspecies, with markedly disjunct distributions. Subsp. *guatemalae* differs from the nominate in having a darker and more violet rump, while subsp. *impatiens* has a more extensive chestnut wing patch. This hummingbird is found mainly at altitudes of 3,300 ft (1,000 m) or lower, but there is some variation between the subspecies, with subsp. *guatemalae* ranging to higher altitudes. The species feeds at all heights in the forest strata, particularly enjoying *Inga* flowers, and is an agile flycatcher. The two more northerly subspecies are reported to be quite common over most of their range. However, subsp. *impatiens* has not been reliably observed in its small known range since the 1950s and may be extinct.

MALE

DISTRIBUTION Subsp. *cyanura* ranges from southern Honduras through east El Salvador to northwest Nicaragua; subsp. *guatemalae* ranges from southeast Mexico to south Guatemala; subsp. *impatiens* is found in northwest and central Costa Rica

HABITAT All kinds of forested habitats, also scrubland and plantations; 0–5,900 ft (0–1,800 m)

SIZE Length: 3½–3⅞ in (9–10 cm). Weight: 4 g

STATUS Least Concern

Steely-vented Hummingbird

This is a shining green hummingbird with darker underparts; a medium-length, straight bill with a black upper mandible and dark-tipped pink lower mandible; and a slightly forked, dark blue tail. The female is similar to the male, but has white edges to the throat feathers. Subsp. *hoffmanni* has a reddish rump, with lighter blue tail feathers; subsp. *warscewiczi* is smaller and has blue-purple undertail coverts; and subsp. *braccata* has purple tips to the rump feathers. The bird feeds at all levels on the nectar of a large range of different herbs, shrubs, and trees, including *Lobelia*, *Hamelia*, *Inga*, and *Tabebuia*; it also catches small insects. The small cup-like nest is built from soft plant material on top of a branch or twig. The total population is unknown, although numbers appear to be on the increase and the species is regarded as common.

DISTRIBUTION Subsp. *saucerottei* is found in northwestern, north-central, and western Colombia; subsp. *hoffmanni* occurs in western Nicaragua and western and central Costa Rica; subsp. *warscewiczi* occurs in northern Colombia and northwestern Venezuela; subsp. *braccata* occurs in western Venezuela

HABITAT Forest edges, semi-open areas, arid areas, scrubland, savanna, plantations, gardens; 0–8,200 ft (0–2,500 m)

SIZE Length: 3⅛–4⅜ in (8–11 cm). Weight: 4–5 g

STATUS Least Concern

FEMALE

Indigo-capped Hummingbird

This is a dark hummingbird with mainly shining yellow-green plumage, shading to orange on the rump. The wings are blackish, the tail is blackish blue, and the crown is dark, shining blue, shading into green towards the nape. The sexes are similar but the female has more restricted blue on the crown. A rare melanistic form has been recorded, and there may be a light blue-crowned subsp. *alfaroana*, although only a single specimen has been recorded, in Costa Rica in 1895, and it is now almost certainly extinct. This species usually feeds at quite high levels and is usually territorial, attempting to drive away all other nectar-feeding birds, though several may assemble at flowering trees. It also flies out over clearings to hawk for insects. The nest is a small cup built on a thin horizontal branch. The species is considered fairly common, though its population trend is unknown.

MALE

DISTRIBUTION Subsp. *cyanifrons* occurs in Northern and central Colombia; subsp. *alfaroana* has only been reported in Costa Rica

HABITAT Wet forest edges, scrub, gardens, plantations; 1,300–9,850 ft (400–3,000 m)

SIZE Length: 2¾–3⅞ in (7–10 cm). Weight: 5 g

STATUS Least Concern

Violet-crowned Hummingbird

DISTRIBUTION Subsp. *violiceps* is found in southwest Mexico; subsp. *ellioti* ranges from southern Arizona and southeast New Mexico to northwest and central Mexico

HABITAT Open woodland and scrub, parks, gardens; 0–7,400 ft (0–2,250 m)

SIZE Length: 3⅞–4⅜ in (10–11 cm). Weight: 5 g

STATUS Least Concern

One of several representatives of the genus *Agyrtria*, which has been subsumed in its entirety into *Amazilia*, this bird looks very similar to the Green-fronted Hummingbird but has a violet-blue rather than green crown. Subsp. *violiceps* has a more rufous tail than subsp. *ellioti*; the latter has more extensive blue tones on the head, spreading from the crown to the upper cheeks. The species feeds at low and high levels, taking nectar from flowers such as Agave, Salvia, and Neobuxbaumia, and is not territorial. The nest may be built near the ground or much higher, usually placed on a horizontal dead twig. After breeding, birds in the north of the range move south for the winter. The species has a rather patchy distribution, especially in the north of its range, but is not thought to be threatened.

MALE

Amazilia fimbriata

Glittering-throated Emerald

This extremely widespread emerald was formerly classed in the now defunct genus *Polyerata* along with several similar species. It is a mainly olive-green hummingbird with a brighter forehead and vivid, glittering green throat patch. The seven described subspecies show slight variation in size and bill length, and subtly different plumage characteristics. The bird is a low-level feeder and mainly a trapliner, visiting all kinds of flowers, including introduced species. The most southerly form, subsp. *tephrocephala*, is a migrant, moving southwards after breeding. The species builds a soft cup nest on a low horizontal branch and incubates its eggs for 14 to 17 days, the chicks fledging another 18 to 22 days later. It is an aggressive emerald and is very common in large parts of its range, with many populations in protected areas. Subsp. *tephrocephala* is considered the most threatened subspecies, due to loss of coastal forest in its range.

DISTRIBUTION Subsp. *fimbriata* occurs from northeast Venezuela to north Brazil; subsp. *elegantissima* occurs in northeast Colombia and north to northwest Venezuela; subsp. *apicalis* occurs in Colombia east of the Andes; subsp. *fluviatilis* occurs in southeast Colombia and east Ecuador; subsp. *laeta* occurs in northeast Peru; subsp. *nigricauda* is found in eastern Bolivia and central Brazil; and subsp. *tephrocephala* occurs on the coast of southeast Brazil

HABITAT Various mainly open habitats, including forest edges, plantations, mangroves, secondary growth, gardens, grassland, and scrub; 0–3,600 ft (0–1,100 m)

SIZE Length: 3⅛–4¾ in (8–12 cm). Weight: 4.5 g

STATUS Least Concern

Sapphire-spangled Emerald

DISTRIBUTION Subsp. *lactea* is found in eastern Brazil; subsp. *zimmeri* occurs in southeast Venezuela; subsp. *bartletti* ranges from east and southeast Peru to north Bolivia

HABITAT Light forest, forest edges, cultivated ground, gardens, plantations; 0–4,600 ft (0–1,400 m)

SIZE Length: 3⅛–4⅜ in (8–11 cm). Weight: 3.5–4 g

STATUS Least Concern

A rather dark hummingbird, this species resembles the Glittering-throated Emerald but its throat patch is deep violet-blue rather than green, and its upperpart coloration is darker and more olive. Subsp. *zimmeri* has a smaller throat patch than the nominate and more white below, and subsp. *bartletti* has a longer bill and wings, less white below, and more gray and brown tones in the plumage. Widely separated, the subspecies also show some behavioral distinctions. For example, the Brazilian subsp. *lactea* lives at much lower altitudes and in the south of its range is migratory, whereas other populations are sedentary. Males of this species are territorial, and both sexes visit a wide range of flower types, including non-native trees and shrubs in gardens. Subsp. *zimmeri* is scarce and localized, found at only a few well-separated sites, but the other forms are quite common.

MALE

Purple-chested Hummingbird

Unusually for its genus, this smart-looking hummingbird shows quite marked sexual dimorphism. The male is predominantly olive green above and below, with a shining green throat bordered below with a horseshoe-shaped, dark violet-blue breast patch. The female has a whitish throat, breast, and belly, the throat and upper breast marked with distinct circular olive-green spots. It is closely related to the Blue-chested Hummingbird and overlaps in range with that species, although it is usually outnumbered by the Blue-chested. It is similar to the Blue-chested in many behavioral respects, although the males are not known to gather at leks as part of courtship. It feeds at low and high levels in the forest strata, visiting *Heliconia*, *Psammisia*, *Tristerix*, and *Costus* flowers, among others. It is fairly common in much of its limited range and occurs in the protected Río Nambi Nature Reserve in Colombia.

DISTRIBUTION West Colombia, northwest Ecuador

HABITAT Rainforest, secondary growth, forest clearings and edges; 0–650 ft (0–200 m)

SIZE Length: 3⅛–3⅞ in (8–10 cm). Weight: 3.5–4 g

STATUS Least Concern

MALE

MALE

Amazilia edward

Snowy-breasted Hummingbird

DISTRIBUTION Subsp. *edward* occurs in the Canal Zone of Panama; subsp. *niveoventer* ranges from Costa Rica to west and central Panama; subsp. *collata* occurs in central Panama; and subsp. *margaritarum* occurs in the northern gulf of Panama and on nearby islands

HABITAT Mainly quite open habitats including savanna and scrub, gardens, mangroves, formerly cultivated land; 0–5,250 ft (0–1,600 m)

SIZE Length: 3⅛–4⅜ in (8–11 cm). Weight: 4.5–5 g

STATUS Least Concern

This species is also known as the Snowy-bellied Hummingbird. It has a clean white belly and rufous undertail, shining green head and breast, and otherwise warm brown upperparts with dark wings. The nominate subspecies has a brown tail, while that of subsp. *niveoventer* is blue and that of subsp. *magaritarum* is bronzy green. The fourth form, subsp. *collata*, has a brown rather than rufous undertail. Where the ranges of subsp. *edward* and subsp. *margaritarum* meet, intergrades occur. This hummingbird is primarily seen at medium heights in the forest strata, either perched on guard over its territory, which it defends from all other hummingbirds, or feeding from flowers of plants such as *Inga*, *Palicourea*, and *Vochysia*. Its habitat preferences vary, but it is inclined to move away from the most arid areas during the dry season. It is a common bird across most of its range.

Green-bellied Hummingbird

A dark, shining green hummingbird, this species has dark wings, a dark violet-blue tail, and a touch of sooty brown on the rump. The sexes are similar, but the females are slightly less bright and have some white fringing to the throat feathers. Recently fledged young birds are drabber still, with gray bellies and more brown on the rump and back. Subsp. *iodura*, which is not always regarded as a valid subspecies and overlaps in range with the nominate, differs in having a more coppery tone to the tail feathers. Like others of its genus this species is quite territorial, and when several birds are attracted to a flowering tree, much vigorous aerial squabbling occurs. It also feeds in the understory. Little is known of its breeding behavior, nor of its population status, although it is fairly common in the Colombian part of its range at least.

JUVENILE

DISTRIBUTION Subsp. *viridigaster* occurs on eastern Andean slopes in Colombia; subsp. *iodura* occurs in the western Venezuelan Andes

HABITAT Forest edges and clearings, secondary growth, plantations; 650–6,900 ft (200–2,100 m)

SIZE Length: 3⅛–3½ in (8–9 cm). Weight: 4.5 g

STATUS Least Concern

Copper-rumped Hummingbird

This shining green-gold hummingbird has
a shining green throat, a reddish-bronze back
and rump, and darker green underparts.
The medium-length, straight bill has a black
upper mandible and a pink base to the lower
mandible. The dark blue tail is forked. The
female is similar to the male, but has a duller
back and rump, and some white on the throat
feathers. Subsp. *erythronotos* is smaller and darker
underneath; subsp. *aliciae* is more reddish above;
subsp. *monticola* is darker above and below;
subsp. *feliciae* has a more golden back; subsp.
caudata has a darker blue tail; and subsp.
caurensis has a grayer rump and a less forked,
dark purple-blue tail. The species feeds at low
to mid-levels on the nectar of a large range of
plants, including *Erythrina*, *Calliandra*,
Palicourea, and *Hibiscus*, and also catches small
insects. It builds a nest from soft plant material
on top of a branch or twig. The total population
and population trend are unknown, although
numbers appear to be stable and the species is
regarded as common.

MALE

DISTRIBUTION Subsp. *tobaci* is found in Tobago; subsp. *erythronotos* occurs in Trinidad; subsp. *aliciae* occurs in Venezuela; subsp. *monticola* occurs in northwestern Venezuela; subsp. *feliciae* occurs in north-central Venezuela; subsp. *caudata* occurs in northeastern Venezuela; subsp. *caurensis* is found in eastern and southeastern Venezuela

HABITAT Forest, forest edges, clearings, semi-open and open areas, savanna, plantations, gardens; 0–6,550 ft (0–2,000 m)

SIZE Length: 3⅛–4⅜ in (8–11 cm). Weight: 4–5 g

STATUS Least Concern

Red-billed Streamertail

The male of this species is a dramatic bird whose second-outermost tail feathers are greatly elongated, making up about half of his body length, and vibrate audibly in flight. The plumage is shining green with a blackish crown, tail, and wings, and the lower crown and cheek feathers are elongated beyond the nape. The bill is red. Females are short-tailed, with green plumage, grayish wings, and white underparts, and have a pinkish-brown bill. A very common bird in all environments, this species is Jamaica's national bird and a popular visitor to gardens, where it takes nectar from flowers of all kinds and utilizes the holes made in larger flowers by flowerpiercers; it may also take sweet sap from holes drilled by woodpeckers. The nest is a delicate cup of plant fibers fixed on a slim twig low in a tree or shrub. Some females rear three broods a year.

FEMALE

DISTRIBUTION Western
and central Jamaica

HABITAT Forest, forest edges,
plantations, gardens; 0–3,300 ft
(0–1,000 m)

SIZE Length: 4⅛–11¾ in
(10.5–30 cm). Weight: 4.5–5 g

STATUS Least Concern

MALE

Golden-tailed Sapphire

This is a spectacular, three-toned hummingbird. The male has a shining, dark blue head, including the throat, with a contrasting green body and bright orange-yellow tail. The female has a green head, concolorous with the body, and a speckled whitish throat. The male of subsp. *josephinae* differs through having a green rather than blue throat. The male is aggressive and somewhat territorial, and quarrelsome gatherings of several males at flowering trees are commonplace; the female tends to be a solitary trapliner. Males form leks of up to ten individuals when seeking a mate, and sing in a complex series of thin, chattering notes. Although not regarded as threatened, this species is declining. The extent of its decline is unclear, but is caused mainly by habitat loss in lowlands across its range.

DISTRIBUTION Subsp. *oenone* occurs in north and east Venezuela, eastern Colombia, and Ecuador, and into far northeast Peru and west Brazil; subsp. *josephinae* ranges from northern Peru to northern Bolivia

HABITAT Humid forest and secondary growth, forest edges and clearings, shady plantations; 0–4,900 ft (0–1,500 m)

SIZE Length: 3¾–3⅞ in (9.5–10 cm). Weight: 4.5 g

STATUS Least Concern

MALE

MALE

Lepidopyga coeruleogularis

Sapphire-throated Hummingbird

This sleek hummingbird is shining metallic green all over except for its dark tail and wings and, in the male, a broad, glistening blue throat patch (the female's throat is white-centered). The tail is quite deeply notched. The nominate has a darker throat patch than subsp. *confinis* (more turquoise-toned) and subsp. *coelina* (lighter blue). Usually a solitary feeder, the species may form small assemblages around flowering trees at the peak of the flowering season. At other times it feeds mainly from low-growing flowers. The nest is built rather low in a shrub on a forked branch, and is a small but deep cup of pale, soft plant fibers. The incubation period is 15 to 16 days, and the fledging period 19 to 22 days. This is a common and adaptable species, using modified habitats and thriving in various protected areas such as the Tayrona National Park in Colombia.

DISTRIBUTION Subsp. *coeruleogularis* occurs in Pacific western Panama; subsp. *confinis* is found on the Caribbean slope in eastern Panama; subsp. *coelina* occurs in the far north of Colombia

HABITAT Mangroves, scrub, light forest, forest edges; 0–350 ft (0–100 m)

SIZE Length: 3⅜–3¾ in (8.5–9.5 cm). Weight: 4–4.5 g

STATUS Least Concern

Sapphire-bellied Hummingbird

A very beautiful hummingbird, this species closely resembles the Sapphire-throated Hummingbird, but in the male the shining blue coloration is darker on the throat and extends to the entire underparts (except for the white undertail). The bill is only very slightly downcurved, and is black above and red below. The female has rarely been reliably observed but is thought to be duller and grayer on the underside. This hummingbird is known to visit *Erythrina fusca* flowers and presumably other local species. With a tiny population and few reliable observations, this is a little-known species in urgent need of study and major conservation help. It is restricted to coastal mangrove swamps, a habitat much damaged by development in its tiny range, particularly a major construction project in the mid-1970s, and there are thought to be no more than 400 individuals left, perhaps far fewer. At least part of the population is protected within Colombia's Salamanca Island Road Park.

MALE

DISTRIBUTION A tiny area of coastal north-central Colombia

HABITAT Mangroves and sometimes adjacent scrubland; 0–350 ft (0–100 m)

SIZE Length: 3½–3¾ in (9–9.5 cm). Weight: 4.5 g

STATUS Critically Endangered

Shining-green Hummingbird

This species differs from the related Sapphire-bellied and Sapphire-throated Hummingbirds in that it has no or very little blue in the plumage. The male of subsp. *goudoti* has an entirely glistening green body plumage with a dark tail and wings, while subsp. *luminosa* is more olive brown. Subsp. *phaeochroa* has a flush of blue on the throat and forehead, and subsp. *zuliae* is small and generally darker than other subspecies. Occasionally visiting *Inga* trees in small but quarrelsome groups, this species also defends territories around low-growing flower clusters along streams and garden borders, and watches for flycatching opportunities from open perches. It builds a small, neat cup nest on a horizontal branch, quite low in a small tree and usually adjoining a clearing. It avoids arid parts of its range during the dry season. It is fairly common and surveys indicate its population is increasing.

DISTRIBUTION Subsp. *goudoti* is found in central Colombia; subsp. *luminosa* occurs in coastal north Colombia; subsp. *phaeochroa* is found in parts of northwest Venezuela; subsp. *zuliae* occurs in the far northeast of Colombia and northwest Venezuela

HABITAT Scrubland, grassland with scattered trees, plantations, gardens, light woodland; 0–5,250 ft (0–1,600 m)

SIZE Length: 3½–3¾ in (9–9.5 cm). Weight: 3.5–4 g

STATUS Least Concern

MALE

Violet-bellied Hummingbird

Superficially similar to the *Lepidopyga* hummingbirds but with a wedge-shaped rather than notched tail, this species has no clear close relationships to other hummingbird genera. The male is shining green, lighter on the head, with a clearly demarcated violet-blue breast and belly, while the female has a pale throat and otherwise green underparts. Subsp. *panamensis* has a darker crown than the nominate, while subsp. *feliciana* is longer-billed and darker-bellied. Rather a little-known species, this hummingbird usually feeds quite low in the forest strata, often near clearings, and males are territorial. Both sexes may join mixed feeding flocks. The bird builds a saddle-style nest on a slim branch in a bush, and in it the female incubates the two eggs for 15 days, then feeds the chicks for 20 to 22 days before they fledge. Though quite inconspicuous, the species is fairly common and occurs in some protected areas.

MALE

DISTRIBUTION Subsp. *julie* is found in north and central Colombia; subsp. *panamensis* occurs in central Panama; subsp. *feliciana* ranges from southwest Ecuador through west Ecuador to northwest Peru

HABITAT Forests and forest clearings and edges, secondary growth; 0–5,900 ft (0–1,800 m)

SIZE Length: 3⅛–3½ in (8–9 cm). Weight: 2.5–3.5 g

STATUS Least Concern

Blue-throated Goldentail

The male Blue-throated Goldentail has dark green-bronze upperparts, turning a bright bronze on the rump, a bright blue-purple throat, and green underparts. The medium-length, straight bill is bright red with a dark tip. The tail feathers are a rich brown-bronze. The female has a mottled blue and gray throat and paler underparts. Subsp. *earina* has darker upperparts and a greener tail. The species feeds on the nectar of many different plants, including *Heliconia*, *Thalia*, *Inga*, and epiphytes, as well as on small insects and spiders. Males sing at leks in the breeding season to attract mates. Birds build a cup-like nest from soft plant fibers and lichen, on top of a branch or twig. The current population is estimated to be fewer than 50,000 individuals, but this is thought to be rising and the species is regarded as fairly common throughout its range.

DISTRIBUTION Subsp. *eliciae* ranges from southeastern Mexico and Belize to southern Costa Rica; subsp. *earina* ranges from western Panama to northwestern Colombia

HABITAT Lowland areas, forest edges, semi-open areas, hedgerows, gardens; 1,650–3,300 ft (500–1,000 m)

SIZE Length: 3⅛–3⅞ in (8–10 cm). Weight: 3.5–4.5 g

STATUS Least Concern

FEMALE

MALE

Hylocharis sapphirina

Rufous-throated Sapphire

The Rufous-throated Sapphire has dark green upperparts, turning a bright bronze on the rump, a bright reddish chin, a shining blue-purple throat and breast, and green underparts. The medium-length, straight bill is bright red with a dark tip. The tail feathers are a rich brown-bronze. The female is similar to the male, but has a paler chin, and a mottled green and gray throat and underparts. The species feeds mostly at low and mid-levels on the nectar of many different flowering plants, including Bromeliaceae, Passifloraceae, Rubiaceae, and Myrtaceae, as well as on small insects and spiders. It builds a cup-like nest from soft plant fibers, leaves, and lichen on top of a branch, usually protected by an overhanging leaf. The current population is unknown, and while numbers appear to be decreasing as a result of deforestation, the species is regarded as fairly common throughout its range.

DISTRIBUTION Eastern Ecuador, northeastern Peru, eastern Colombia, southern Venezuela, and the Guianas to north and central Brazil and northeastern Bolivia; separate populations in southeastern Brazil and from eastern Paraguay to northeastern Argentina

HABITAT Lowland areas, forest edges, semi-open areas, clearings, plantations, gardens; 650–6,550 ft (200–2,000 m)

SIZE Length: 3⅛–3⅞ in (8–10 cm). Weight: 3.5–5 g

STATUS Least Concern

White-chinned Sapphire

Very similar to the Golden-tailed Sapphire, with shining green upperparts that grade to reddish bronze on the rump, a shining blue-purple forehead, throat, and breast with mottled green edges, and gray underparts, the White-chinned Sapphire can be differentiated from that species by its throat feathers, which are white underneath. The medium-length, straight bill is bright red with a dark tip, and the tail feathers are dark blue. The female is similar to the male, but is paler and has a dull green head and back. Subsp. *viridiventris* is darker underneath than the nominate and has a shorter bill; subsp.

rostrata is larger; subsp. *conversa* is paler underneath and has a shorter bill; and subsp. *griseiventris* is more blue-green on its head and throat and grayer underneath. The species feeds on the nectar of many different plants, including Bromeliaceae, Rubiaceae, and Myrtaceae, as well as on small insects and spiders. It builds a cup-like nest from soft plant material on top of a branch. The current population is unknown, but numbers appear to be decreasing as a result of deforestation. Despite this, the White-chinned Sapphire is regarded as fairly common throughout its range.

MALE

DISTRIBUTION Subsp. *cyanus* occurs in eastern Brazil; subsp. *viridiventris* ranges from northern and eastern Colombia to western and southern Venezuela, the Guianas, and northern Brazil; subsp. *rostrata* occurs in eastern Peru, northern Bolivia, and western Brazil; subsp. *conversa* is found in eastern Bolivia, northern Paraguay, and southwestern Brazil; subsp. *griseiventris* occurs in southeastern Brazil and northeastern Argentina

HABITAT Lowland areas, forest edges, semi-open areas, clearings, plantations, gardens; 0–3,300 ft (0–1,000 m)

SIZE Length: 3⅛–3⅞ in (8–10 cm). Weight: 3–5 g

STATUS Least Concern

Gilded Hummingbird

The Gilded Hummingbird is shining, light gold-green above and below, with a small orange patch beneath the bill and a very small white spot behind each eye. The medium-length, straight bill is red with a dark tip, and the tail feathers are bronze-green. The female is similar to the male, but is paler overall and dull gray underneath. The species feeds on the nectar of many different plants at all levels, including *Inga*, *Citrus*, *Hibiscus*, *Abutilon*, and *Salvia*; it also hawks for small insects and spiders. The cup-like nest is built from plant material on top of an exposed branch. The current population is unknown, but numbers appear to be increasing and the species is regarded as fairly common throughout its range.

MALE

DISTRIBUTION Northeastern and south-central Bolivia, Paraguay, southeastern and central Brazil, Uruguay, northern Argentina

HABITAT Lowland areas, forest edges, semi-open areas, clearings, savanna, plantations, gardens; 0–3,300 ft (0–1,000 m)

SIZE Length: 3½–3⅞ in (8–10 cm). Weight: 3–5 g

STATUS Least Concern

Blue-headed Sapphire

The male of this species is shining green with a blue crown, cheeks, and tail, and has a red bill tipped with black. The female has a dark bill and is a slightly duller shade of green, with a pale underside that is heavily spotted with green. This species will feed from many flowering trees and shrubs, and will sometimes gather in quite large numbers around trees that are in flower. However, it is intolerant of close approach by conspecifics and each bird attempts to defend a (small) feeding territory. Little is known of its breeding behavior. Its patterns of occurrence are rather unpredictable, suggesting that it undertakes quite marked seasonal movements. Although it seems willing to use modified habitats to some extent, parts of its range have been completely converted to crop monocultures and the species is thought to be declining.

DISTRIBUTION Western Colombia, north Ecuador

HABITAT Dry woodland and scrub, plantations, forest edges; 1,650–8,550 ft (500–2,600 m)

SIZE Length: 3⅞–4⅜ in (10–11 cm). Weight: 6–7 g

STATUS Least Concern

MALE

Xantus's Hummingbird

This species and the very closely related White-eared Hummingbird show considerable similarity to the mountain-gems (genus *Lampornis*), and are sometimes placed separately in the genus *Basilinna*. This is a shining green hummingbird with a rich rufous belly and tail sides. The male's face is mostly black, while in the female the orange-rufous coloration spreads to the throat and forehead. Both sexes have a bold white stripe behind the eye. The species feeds from various local flower species, including *Calliandra peninsularis*, *Fouquiera diguetii*, and *Mirabilis jalapa*, and is particularly dependent on *Arbutus peninsularis* in late winter. It also picks insects from leaves and bark. The nest is a sturdy cup of plant fibers, fixed with cobwebs low in a tree and often overhanging water; incubation takes 15 to 16 days and fledging 20 to 22 days. Though normally sedentary, Xantus's Hummingbird has wandered to California on occasion.

MALE

DISTRIBUTION Southern Baja California

HABITAT Deciduous and mixed pine–oak forest, sometimes cultivated land, dense scrub with cacti, usually near running water; 5,900–7,200 ft (1,800–2,200 m)

SIZE Length: 3⅛–3½ in (8–9 cm). Weight: 3–4 g

STATUS Least Concern

White-eared Hummingbird

Very like Xantus's Hummingbird in general appearance and pattern, the White-eared Hummingbird differs in that it lacks extensive rufous tones. The male is rich, shining green with a dark, shining violet-blue face, shading to black on the cheeks, and the female is green above and white below with green scaling on the throat and breast. Both sexes have a white stripe behind the eye. Subsp. *borealis* has bronze-toned upperparts, as does subsp. *pygmaea*, which is also the smallest of the three subspecies. Males of this species are territorial around good nectar sources and will drive away even larger species, but will tolerate each other when singing at leks to attract females. Courtship includes chasing flights and the male hovering directly in front of the perched female. The nest, built in a shrub or small tree, is well concealed and heavily camouflaged. This bird is common over much of its range.

DISTRIBUTION Subsp. *leucotis* ranges from central and southern Mexico to Guatemala; subsp. *borealis* is found in north Mexico; subsp. *pygmaea* ranges from El Salvador through Honduras and Nicaragua

HABITAT Pine and mixed pine–oak forest and forest clearings; 3,950–11,500 ft (1,200–3,500 m)

SIZE Length: 3½–3⅞ in (9–10 cm). Weight: 3–3.5 g

STATUS Least Concern

MALE

Scissor-tailed Hummingbird

The male of this species—the only member of its genus—is unmistakable, with a very long, deeply forked tail that contributes to about half of his total body length. The female in contrast has a much shorter tail. The male's plumage is dark, shining green, with a highly iridescent throat and breast patch; the female has pale, dark-spotted underparts and a yellow-orange tail tip and tail sides. A territorial feeder, the bird visits bromeliads and *Heliconia* and *Costus* flowers, and it also takes insects. Its status has changed several times since 1994: It was assessed initially as Critically Endangered and then as Vulnerable, before being changed to Endangered in 2008. It is particularly at risk because of its small population and tiny range— just 5.7 square miles (14.8 km^2) in northeast Venezuela of prime habitat remain, though the species does tolerate secondary growth to some degree.

DISTRIBUTION Paria Peninsula, northeast Venezuela

HABITAT Humid, mossy cloud forest, occasionally forest edges; 2,950–3,950 ft (900–1,200 m)

SIZE Length: 4¾–7½ in (12–19 cm). Weight: 6.5–8 g

STATUS Endangered

MALE

Magnificent Hummingbird

This aptly named hummingbird was formerly known as Rivoli's Hummingbird in honor of nineteenth-century French amateur ornithologist François Victor Masséna, the second Duke of Rivoli. Females are similar to female Blue-throated Hummingbirds, though they have more green and less white in the tail. Young males are intermediate between females and adult males, with dark, scaly-looking underparts and patches of bright iridescence in the gorget and crown. Northern birds of dry forests are trapliners, foraging at many small flower patches over the course of a day, while southern birds of humid forests tend to be more territorial. The northernmost populations are migratory, though a few individuals in the southwestern United States frequent feeding stations all year-round. Populations in Costa Rica and Panama are larger and paler than northern birds and are sometimes considered a separate species, the Admirable Hummingbird (*Eugenes spectabilis*).

DISTRIBUTION Subsp. *fulgens* occurs in the southwestern United States, Mexico, Guatemala, El Salvador, and Honduras; subsp. *spectabilis* occurs in Costa Rica and extreme western Panama

HABITAT Mountain conifer, oak, and mixed forest; 4,900–9,850 ft (1,500–3,000 m), often lower during migration and in winter

SIZE Length: 4¾–5½ in (12–14 cm). Weight: 5.5–9 g

STATUS Least Concern

MALE

Fiery-throated Hummingbird

This is a dark and very highly iridescent hummingbird with a vivid red-centered orange throat patch that contrasts with the dark blue forehead and upper breast. The plumage is otherwise shining green. Subsp. *eisenmanni* has more extensive blue coloration, which reaches the belly, and a shorter bill than the nominate. Males and females are similar to each other in their appearance, and also in their aggressive, territorial habits. The species visits a wide range of flowers, primarily high in the forest canopy, piercing the corollas of longer flowers. The nest is a substantial cup of cobwebs and soft plant down, built at the tip of a dangling root or stem and decorated with lichen and moss. After the chicks fledge, many birds move to lower elevations. This is a common bird across much of its restricted range, and can be seen in several protected areas.

DISTRIBUTION Subsp. *insignis* is found from north-central Costa Rica south to west Panama; subsp. *eisenmanni* is found in north Costa Rica

HABITAT Cloud forest, high-altitude scrub and pasture with trees; 5,250–10,500 ft (1,600–3,200 m)

SIZE Length: 4⅛–4⅜ in (10.5–11 cm). Weight: 5–6 g

STATUS Least Concern

MALE

Long-billed Starthroat

The Long-billed Starthroat comprises three subspecies that are widespread in the humid tropics but seldom common. The species is similar to the Plain-capped, whose range it overlaps along parts of the Pacific coast of Central America. This hummingbird is a long-winged, graceful flier, often gliding short distances on set wings during extended hawking flights high over open areas. The sexes are similar, though the adult female has little or no blue on the crown and a duller red gorget. The nest is similar to those of the mangoes, often straddling a high, open tree branch and camouflaged with chips of gray-green lichen. Young birds leave the nest with dusky gorgets and solid-green crowns. The species is mainly a trapliner, though males may defend rich nectar sources such as flowering coralbean trees (*Erythrina*). Though the species is not migratory in the traditional sense, annual changes in abundance and sightings outside its normal range suggest some seasonal movements.

DISTRIBUTION Subsp. *longirostris* ranges from southwestern and eastern Costa Rica to eastern Bolivia and central and northeastern Brazil, and also includes a separate population on Trinidad; subsp. *albicrissa* occurs in western Ecuador and northwestern Peru; subsp. *pallidiceps* ranges from southern Mexico to Nicaragua

HABITAT Humid forest canopy and edges, open woodland, riparian galleries, savanna, pastures, plantations; 0–4,900 ft (0–1,500 m)

SIZE Length: 3⅞–4¾ in (10–12 cm). Weight: 6.5–7 g

STATUS Least Concern

FEMALE

Stripe-breasted Starthroat

The strikingly marked Stripe-breasted Starthroat is virtually unknown outside Brazil. Females are similar to those of northern starthroat species, with broad white mustaches, gray underparts with a white midline stripe, a narrow gorget of dark green feathers with pale edges, and white tips on the outer tail feathers. Both sexes have jagged white patches on the lower back and silky white flank tufts that protrude from beneath the wings at rest. As in the closely related Blue-tufted Starthroat, adult males adopt a dull eclipse plumage in the non-breeding season. The species name *squamosus*, which is Latin for "scaly," may refer to the dark-centered gorget feathers of females and eclipse males. Birds take nectar from a wide variety of native and cultivated flowers, including *Hibiscus*, Royal Poinciana (*Delonix regia*), and large bromeliads (*Billbergia*), and readily accept sugar water from feeders.

FEMALE

MALE

DISTRIBUTION Eastern Brazil

HABITAT Tropical and subtropical forest, grassland, savanna, pastures; 0–2,600 ft (0–800 m)

SIZE Length: 4⅜–4⅞ in (11–12.5 cm). Weight: 5–6.5 g

STATUS Least Concern

MALE

Blue-tufted Starthroat

The Blue-tufted Starthroat is among the few hummingbirds known to wear a less colorful eclipse plumage for part of each year. At the end of the breeding season, adult males retain most of their distinctive markings but replace their gaudy magenta and blue gorgets with dull white feathers. The female's plumage is understated year-round, with a plain white throat narrowing to a faint white midline stripe on pale gray underparts. Both sexes have the prominent white flank tufts found in other starthroats but lack a jagged white patch on the lower back. Birds hawk insects in flight and use their long bills to extract nectar from a variety of flowers, including the tree cactus *Opuntia quimilo*, coralbeans (*Erythrina*), and *Canna indica*. Rare sightings in northeastern Ecuador and southeastern Colombia may indicate some northward migration during the austral winter.

DISTRIBUTION Central and southern Brazil, southern Bolivia, western and central Paraguay, northern Uruguay, northern Argentina

HABITAT Dry savanna, grassland, tropical and subtropical lowland forest edges; 0–3,300 ft (0–1,000 m)

SIZE Length: 4⅜–5⅛ in (11–13 cm). Weight: 5–6.5 g

STATUS Least Concern

Green-throated Mountain-gem

The Green-throated Mountain-gem has a distinctive white eye stripe running down from behind the eye, shining green upperparts, and a dark bronze rump. The white throat is mottled with light blue-green in the male and plain in the female. The dull, dark brown tail is slightly forked. The medium-length, straight bill is black. Subsp. *amadoni* differs from the nominate in having a darker, greener rump; subsp. *ovandensis* has less green on its underparts; and subsp. *nubivagus* is dark green above, with a bright orange-bronze rump. Very little is understood about the species' favored food plants or breeding behavior, but it is known to feed at all levels on the nectar of many plant species, as well as on insects and small spiders. The current population is thought to be in decline, but the bird is regarded as locally common within its range.

DISTRIBUTION Subsp. *viridipallens* occurs in Guatemala, northern El Salvador, and western Honduras; subsp. *amadoni* occurs in southern Mexico; subsp. *ovandensis* is found in southern Mexico and northwestern Guatemala; subsp. *nubivagus* occurs in El Salvador

HABITAT Forests, forest edges; 2,300–9,850 ft (700–3,000 m)

SIZE Length: 3⅞–4¾ in (10–12 cm). Weight: 4.5–6 g

STATUS Least Concern

MALE

Green-breasted Mountain-gem

Similar to the Green-throated Mountain-gem, this medium-sized hummingbird has a distinctive white eye stripe running down from behind the eye, dull green cheeks, and a shining green throat patch. It also has shining green upperparts with a dark bronze rump, and is darker green underneath. The dull, dark brown tail is slightly forked and the medium-length black bill is straight. The female is similar to the male, but has a paler, creamy throat patch. Very little is understood about the species' favored food plants or breeding behavior, but it is known to feed at all levels on the nectar of many different plants, as well as on insects and small spiders. The total population and population trend are unknown, although numbers appear to be stable, and the species is regarded as fairly common throughout its range.

DISTRIBUTION Eastern Honduras, north-central Nicaragua

HABITAT Mountainous areas, forest, forest edges, brush; 2,300–8,200 ft (700–2,500 m)

SIZE Length: 3⅞–4¾ in (10–12 cm). Weight: 4.5–6 g

STATUS Least Concern

MALE

Amethyst-throated Hummingbird

This hummingbird is rather long-billed, and the male has a dark, shining throat patch; the throat of the female is light orange. Otherwise, the plumage is light, shining yellow-green above, turning bronzy on the rump, and gray below, with dark ear coverts and a white line behind the eye. The male of subsp. *amethystinus* has a reddish-violet throat patch, while that of the male subsp. *margaritae* is deep violet-blue. Subsp. *circumventus* is generally paler with a pinkish throat patch; subsp. *salvini* is a smaller and darker form; and subsp. *nobilis* is smaller and darker still. This is a low-level feeder and a habitual trapliner, visiting *Salvia, Fuchsia,* and similar flower types as well as taking many insects. The nest, a cup made of moss, is built rather low on hanging twigs. Though still fairly common and widespread, this species is declining because of deforestation in its range, and its distribution is consequently becoming more patchy.

DISTRIBUTION Subsp. *amethystinus* is found in west, central, and eastern Mexico; subsp. *margaritae* occurs in southwest Mexico; subsp. *circumventus* occurs in south Mexico; subsp. *salvini* ranges from south Mexico to Guatemala and El Salvador; subsp. *nobilis* occurs in Honduras

HABITAT Humid forest and forest edges; 2,950–9,850 ft (900–3,000 m)

SIZE Length: 4½–4⅞ in (11.5–12.5 cm). Weight: 5–8 g

STATUS Least Concern

MALE

Blue-throated Hummingbird

In the lush canyons of Mexico's Sierra Madre and the "sky islands" of the southwestern United States, male Blue-throated Hummingbirds defend territories with a squeaky song, like the turning of a rusty wheel, and flashes of their boldly marked tails. Subsp. *bessophilus* is duller and paler and subsp. *phasmorus* is greener above. The sexes are similar, except that the smaller, longer-billed female lacks the sky-blue gorget. Females build moss-covered nests in the shelter of undercut banks, rock ledges, caves, and human structures, adding material with each successive nesting. Males defend patches of flowers such as Cardinal Flower (*Lobelia cardinalis*), red and pink penstemons (*Penstemon*), columbines (*Aquilegia*), and sages (*Salvia*). Both sexes are enthusiastic predators, flycatching over water or above the tree canopy and gleaning invertebrates from foliage and bark. Northern populations retreat south in winter, though a few individuals are year-round residents around feeders in the southwestern United States.

MALE

DISTRIBUTION Subsp. *clemenciae* ranges from northeastern to southern Mexico; subsp. *phasmorus* occurs in southwestern Texas; subsp. *bessophilus* ranges from southern New Mexico and southeastern Arizona to northwestern Mexico

HABITAT Mountain conifer, broadleaf, and mixed forest edges, moist wooded canyons; 3,300–12,800 ft (1,000–3,900 m)

SIZE Length: 4¾–5⅜ in (12–13.5 cm). Weight: 5.5–9 g

STATUS Least Concern

MALE

Lampornis hemileucus

White-bellied Mountain-gem

The White-bellied Mountain-gem has a shining green head and cheeks, a distinctive white eye stripe running down from behind the eye, and a shining purple throat patch. It has shining green upperparts, and white underparts with green mottling along its sides. The tail feathers are a dull, dark brown. The comparatively short black bill is straight. The female is similar to the male, but has a duller head and lacks a throat patch. Males are aggressive and often territorial. The species feeds on the nectar of many different plants and trees at all levels, including *Inga*, *Quararibea*, Acanthaceae, and Ericaceae; it also catches small insects and spiders. Very little is known about the species' nesting behavior. The total population and population trend are unknown, although numbers appear to be relatively stable and the species is regarded as fairly common throughout its range.

DISTRIBUTION Northern Costa Rica to western Panama

HABITAT Wet forest, forest edges, brush; 2,300–4,600 ft (700–1,400 m)

SIZE Length: 3⅞–4¾ in (10–12 cm). Weight: 5–7 g

STATUS Least Concern

Purple-throated Mountain-gem

This species is a striking bird, both sexes being colorful and very distinct. The male is glossy green with a light turquoise forehead and deep violet throat patch, and has a dark tail and wings. The female is shining green above and rich cinnamon below. Both sexes have a fine white stripe behind the eye. Males usually feed near the canopy, while females feed at lower levels. Both sexes hunt insects, males more often by hawking and females by gleaning. The female nests rather low in the understory, usually next to a clearing of some kind, and camouflages the nest cup with outer layers of moss and lichen. This species has a small range that has suffered some deforestation, but although it is probably declining it is currently not considered to be of conservation concern.

FEMALE

MALE

DISTRIBUTION Subsp. *calolaemus* occurs in central Costa Rica; subsp. *pectoralis* is found in southwestern Nicaragua and northwestern Costa Rica; subsp. *homogenes* occurs in western Panama

HABITAT Wet forest and forest edges; 3,950–8,200 ft (1,200–2,500 m)

SIZE Length: 3⅞–4½ in (10–11.5 cm). Weight: 4.5–6 g

STATUS Least Concern

MALE

White-throated Mountain-gem

The White-throated Mountain-gem has a shining blue head, a distinctive white eye stripe running down from behind the eye, and a bright white throat patch above a shining blue-green breast. It has shining green-bronze upperparts and dull green-gray underparts. The medium-length black bill is straight. The female is similar to the male, but has a duller head, lacks the throat patch, and is pale orange underneath. Subsp. *castaneoventris* has a blue-black tail, while that of subsp. *cinereicauda* is gray. The species forages at lower levels in more open areas, feeding on the nectar of many different plants and trees, including *Fuchsia*, *Cavendishia*, *Satyria*, *Inga*, Acanthaceae, and Ericaceae; it also catches small insects and spiders. Males are aggressive and often territorial. The cup-like nest is built from plant material and usually hidden in the understory in a bush or tree. The total population and population trend are unknown, although numbers appear to be relatively stable and the species is regarded as fairly common throughout its range.

DISTRIBUTION Subsp. *castaneoventris* occurs in western Panama; subsp. *cinereicauda* is found in southern Costa Rica

HABITAT Mountainous areas, wet forest, forest edges, brush, semi-open areas; 3,300–6,550 ft (1,000–2,000 m)

SIZE Length: 3⅞–4¾ in (10–12 cm). Weight: 4–7 g

STATUS Least Concern

Garnet-throated Hummingbird

This strikingly colorful hummingbird has a relatively short, straight bill. The male is glistening green, brightest on the crown, with a reddish-orange throat, violet breast, and black belly. The female's underside is gray, and she has a vertical white stripe behind the eye. Both sexes have bright rufous edges to the wing feathers, and dark cheeks. Males are strongly territorial, and both sexes feed from flowering shrubs and trees such as *Inga* and *Erythrina* at low and medium heights in the strata. The nest is a well-insulated, sizable cup with a soft lining, made from moss, pine needles, and other leaf fragments, and is often fixed to exposed roots on dried-up stream banks. The species nests at the lower end of its altitudinal range; after breeding, the birds move to higher altitudes. It is quite common and will use modified habitats.

MALE

DISTRIBUTION Central and south Mexico, south to Guatemala, El Salvador, and Honduras

HABITAT Cloud forest, pine–oak woodland, other tropical forest habitats, scrubland; 3,950–9,850 ft (1,200–3,000 m)

SIZE Length: 4¾–4⅞ in (12–12.4 cm). Weight: 5.5–7 g

STATUS Least Concern

Chilean Woodstar

This tiny bird is sometimes placed in the genus *Myrtis* with the Purple-collared Woodstar, and as with that species the male has a dark, forked tail while the female's is rounded with white tips to all but the central feathers. The male is shining yellow-green above and white below with a violet throat, while the female's underside is wholly pale buff. The bill is short and straight. The species has been observed feeding from a wide range of native and introduced trees and shrubs, but its ongoing decline (leading to a change in its conservation status from Vulnerable to Endangered in 2000) may be due to loss of preferred native plant species. Competition with the Peruvian Sheartail, a territorial hummingbird that has recently spread into the Chilean Woodstar's range, is another possible but as yet unproven factor in the latter's decline.

DISTRIBUTION Southern Peru and northern Chile

HABITAT Scrubland in desert river valleys, gardens; 650–2,300 ft (200–700 m), rarely higher

SIZE Length: 3–3⅛ in (7.5–8 cm). Weight: 2.5 g

STATUS Endangered

MALE

MALE

Rhodopis vesper

Oasis Hummingbird

This is a relatively large hummingbird with a very long, slightly downcurved bill and, in the male, a long, forked tail. It is olive green above and white below, with a white eyebrow and orange rump. The male additionally has a large, blue-edged, iridescent purple throat patch, which becomes duller outside the breeding season. There are three subspecies—subspp. *vesper*, *koepckeae*, and *atacamensis*—which vary slightly in bill and tail length. The species feeds at trees and shrubs of the families Malvaceae and Fabaceae. The nest is a small cup suspended from a thin tree branch well above the ground; incubation and fledging periods are 15 to 16 days and 22 to 24 days, respectively. The Oasis Hummingbird is a common and adaptable species with a stable population, its future considerably brighter now that trapping for the bird trade has largely ceased.

DISTRIBUTION Subsp. *vesper* occurs in Peru and north Chile; subsp. *koepckeae* occurs in northwest Peru; subsp. *atacamensis* occurs in north Chile

HABITAT Dry grassland, oases, gardens, thinly vegetated mountainsides; 0–9,850 ft (0–3,000 m)

SIZE Length: 5⅛–5⅜ in (13–13.5 cm). Weight: 3.5–4 g

STATUS Least Concern

Peruvian Sheartail

This is a very small, short-billed hummingbird, though the male's greatly elongated central tail feathers make its total length almost double that of the short-tailed female. The male also sports a glittering, blue-edged violet throat patch. Both sexes are otherwise green above and pale below. The species feeds from flowers of plants such as *Russelia* and *Cordia*, usually 6 to 10 feet (2 to 3 m) above ground. Male Peruvian Sheartails are aggressively territorial, defending a favorite flowering cactus, shrub, or tree from all-comers, and their energetic activity sometimes leads to breakage of the long tail feathers. The female builds a tiny, soft cup nest in the fork of a vertical branch, and incubates the clutch of two eggs for 15 to 16 days. The chicks fledge at 20 to 23 days. A widespread species whose habitat is not threatened, this hummingbird's population is stable.

MALE

DISTRIBUTION Western Peru, northern Chile

HABITAT Open, semiarid coastal areas and mountain slopes; 0–7,850 ft (0–2,400 m)

SIZE Length: 2¾–5⅛ in (7–13 cm). Weight: 2–2.5 g

STATUS Least Concern

White-bellied Woodstar

Like other woodstars,
this is a small but stocky,
compact hummingbird. It has
a slightly ragged-looking forked
tail and a straight, needle-like bill.
The male is shining, dark green above
with a reddish-violet throat, white belly,
and green flanks, while the female is duller
with a pale buff throat. Birds are usually seen
alone, and feed from flowers at all levels from
ground to canopy, hovering in a rather moth-
like manner. Particularly rich clusters of flowers
may attract small groups, and the species also
takes insects in flight. It is not territorial and
its slow, insect-like flight tends not to draw the
attention of other, more aggressive species. It
seems to be generally rather uncommon, though
it is easily overlooked. Overall numbers are
thought to be stable, though the species has a
very patchy distribution, so smaller and more
isolated populations may be at risk.

MALE

DISTRIBUTION The central and
eastern Andes in Colombia
through Ecuador and Peru to
central Bolivia

HABITAT Humid forest edges,
pastures, cultivated land;
4,900–9,200 ft (1,500–2,800 m)

SIZE Length: 3⅜ in (8.5 cm).
Weight: 4 g

STATUS Least Concern

Little Woodstar

The male of this tiny, round-bodied, large-headed woodstar has a striking red throat and spiky-tipped tail feathers. Both sexes have otherwise buff underparts and mid-green upperparts (more brown-toned in the female). The species' resemblance to a bumblebee when flying is highlighted in its scientific name (*Bombus* is the genus name of bumblebees). It is very inconspicuous, both to human observers and to territorial hummingbirds, because of its small size and mode of flight, but may be seen hovering at flowers at low levels within the forest, or perching on a high bare twig. The flowers it visits include *Agave*, *Inga*, *Palicourea*, and *Psammisia*. The species is uncommon and declining, with a distribution that is probably highly fragmented, and its habitat is under threat from deforestation. This is reflected in its conservation status being classed as Vulnerable. A full population assessment is needed to plan effective conservation.

MALE

FEMALE

DISTRIBUTION The Andes in the far southwest of Colombia through west Ecuador to north Peru, and east Ecuador and north-central Peru

HABITAT Deciduous forest in between humid and semihumid regions; 0–9,850 ft (0–3,000 m)

SIZE Length: 2⅜–2¾ in (6–7 cm). Weight: unknown

STATUS Vulnerable

Gorgeted Woodstar

The male of this pretty woodstar has a red-violet throat patch that spreads onto the breast sides, with a white breast band below it. The plumage is otherwise dark blue-green above and below. The female has bronze-green upperparts and is buff below. Aside from the nominate there is one subspecies, the darker, shorter-tailed subsp. *cleavesi*. A typical woodstar in its habits, this species usually forages alone and can often successfully enter territories of more aggressive birds because of its inconspicuous flight style. It usually feeds from mid-levels in the forest up to the canopy, and takes many insects. The nest is built on a horizontal branch and is composed of soft, fluffy plant seeds, with an outer layer of lichen for camouflage. There is evidence of some altitudinal dispersal after breeding. This is a common bird in some parts of its range and its population, though highly fragmented, is stable.

MALE

DISTRIBUTION Subsp. *heliodor* occurs in the Andes of Venezuela, Colombia, and west Ecuador; subsp. *cleavesi* occurs in the Andes of northeast Ecuador

HABITAT Wooded areas, including plantations, forest edges, and large gardens; 3,950–9,850 ft (1,200–3,000 m)

SIZE Length: 2¼–2½ in (5.8–6.4 cm). Weight: unknown

STATUS Least Concern

Santa Marta Woodstar

Formerly regarded as a subspecies of the Gorgeted Woodstar, this hummingbird closely resembles that species, the male having a flared throat patch of shining, light reddish violet, a white breast patch, and an otherwise green plumage. The female is light yellow-buff below, and a brighter green above than the female Gorgeted Woodstar. The species feeds on nectar from a range of flowering plants and also hawks for insects. Its restricted distribution in an area with very difficult access means that it has been little studied, though its general behavior is likely to be very similar to other closely related species. Although it has a small range, it is reported to be fairly common where it occurs, and its apparent readiness to accept heavily modified habitats such as coffee plantations would suggest that it can adapt to the changes brought about by the deforestation ongoing in its range.

FEMALE

DISTRIBUTION Sierra Nevada de Santa Marta, northeast Colombia

HABITAT Montane forest edges, coffee plantations, other woodland; 2,700–6,550 ft (825–2,000 m)

SIZE Length: 2¾ in (7 cm). Weight: unknown

STATUS Least Concern

MALE

Esmeraldas Woodstar

Both sexes of this woodstar have mostly white underparts and green flanks. The male's broad throat patch is dark purple, his upperparts are brighter green than the female's, and his tail feathers are elongated and spiky. A canopy-feeder, this species has the same unobtrusive habits and weaving, bumblebee-like flight as its congeners. Plants known to be visited include *Inga* and *Muntingia*. Post-breeding, it moves to much higher altitudes. The forest used by the Esmeraldas Woodstar is extremely threatened and large areas have already been destroyed in the last few decades, with the result that the species' population is now very small (no more than 1,500 individuals), highly fragmented, and rapidly declining. Surveys in the early twenty-first century have identified 15 breeding sites; further surveying and robust habitat protection are needed.

DISTRIBUTION The Pacific slopes of the Andes in western Ecuador

HABITAT Moist lowland forest edges, secondary growth; 0–6,250 ft (0–1,900 m)

SIZE Length: 2⅜–2¾ in (6–7 cm). Weight: unknown

STATUS Endangered

Short-tailed Woodstar

A small, compact, very short-tailed hummingbird, this species has a slightly downcurved bill. Both sexes have green upperparts; the female is warm buff below, while the male's underside is white with a relatively small, bright purple throat patch and some green on the flanks. Short-tailed Woodstars can be seen feeding at flowers 3 to 10 feet (1 to 3 m) above ground level. When the bird perches, the wingtips extend beyond the tail tip. The nest is a tiny downy cup, built on a branch fork in a shrub; sometimes several females construct their nests in close proximity. Incubation lasts 15 to 16 days, and while the chicks fledge at 20 to 23 days they may return to the nest after the first flight. This adaptable species is common throughout its range, and accepts modified habitats without difficulty. Its population trend is stable.

DISTRIBUTION West Ecuador to northwest Peru

HABITAT Open, dry coastal scrub, gardens, valleys; 0–650 ft (0–200 m)

SIZE Length: 2⅜–2½ in (6–6.5 cm). Weight: 2.5–3.5 g

STATUS Least Concern

MALE

Bahama Woodstar

The regal Bahama Woodstar occurs throughout the Lucayan Archipelago, a chain of low-lying islands that includes the Bahamas, and Turks and Caicos Islands. Of these, the southernmost islands, Great and Little Inagua, are home to a distinct subspecies, *lyrura*, in which the rose-purple of the male's gorget extends to the forecrown. The males are among the few hummingbirds to adopt an "eclipse" plumage after breeding, replacing their iridescent gorgets with drab gray feathers. Females are creamy white below with a rusty wash on the sides and long, forked tails boldly banded with rufous, green, black, and creamy white to pale rufous. Bahama Woodstars do not compete well with the larger and more aggressive Cuban Emerald and are less common on islands where both species occur. Though the species is sedentary, occasional wanderers have found their way to the mainland: There are several records from Florida and a single record from Pennsylvania.

DISTRIBUTION Subsp. *evelynae* occurs on islands of the Bahamas except the Inaguas, and on Turks and Caicos; subsp. *lyrura* is restricted to Great and Little Inagua

HABITAT Woodland, scrub, mixed pine forest, forest edges, and clearings, gardens; 0–200 ft (0–60 m)

SIZE 3½–3¾ in (9–9.5 cm), Weight: 2.5–3.5 g

STATUS Least Concern

FEMALE
WITH CHICKS

Magenta-throated Woodstar

The male Magenta-throated Woodstar has shining green-bronze upperparts, a bright, shining purple throat above a broad white collar, a small white spot behind each eye, and orange-red underparts with mottled green either side. The tail feathers are a dark bronzy brown, and the short, straight bill is black. The female's throat is plain cream and the collar is more mottled, and her tail feathers are shorter. The flight "hum" is much louder than in most hummingbird species. Males are aggressive and give territorial displays by swooping back and forth while making loud, mechanical bleating noises, followed by manakin-like snaps. The species feeds at all heights on the nectar of many different shrubs, epiphytes, and trees, including *Quararibea*, *Inga*, *Lantana*, *Lobelia*, and Ericaceae; it also catches small insects and spiders. The nesting behavior is unknown. The total population and population trend are also unknown, although numbers appear to be relatively stable. The species is regarded as uncommon and is patchily distributed throughout its range.

FEMALE

MALE

DISTRIBUTION Northern Costa
Rica, western Panama

HABITAT Mountainous areas,
forest edges, clearings, scrubland,
semi-open slopes; 1,650–6,550 ft
(500–2,000 m)

SIZE Length: 2¾–3⅞ in (7–10 cm).
Weight: 3–4 g

STATUS Least Concern

MALE

Calliphlox mitchellii

Purple-throated Woodstar

The male of this species, which is also known as Mitchell's Woodstar, is dark green-bronze above with a dark purple throat that looks almost black in some light. It has a white breast and dark, mottled green-bronze underparts that grade to orange-red on the sides and towards the feet. There is a small white spot behind each eye, and the short, straight bill is black. The forked tail feathers are brown. The female's throat is plain cream, mottled at the side, and she has a white line that runs from behind the eye down to the white collar. The female's outer tail feathers are not as long as the male's, and are pale orange with a dark band. The flight "hum" is much louder than in most other hummingbird species, and the male's diving territorial displays are accompanied by loud, mechanical noises. The species feeds mainly in the canopy on the nectar of many different plants and trees, including *Inga* and *Cordia*; it also catches small insects. The small cup-like nest is built from fine plant material on a broad branch high in a tall tree. The total population and population trend are unknown, although numbers appear to be stable. The species is regarded as uncommon and is patchily distributed throughout its range.

DISTRIBUTION Eastern Panama through western Colombia and western Ecuador

HABITAT Forest, forest edges; 0–8,200 ft (0–2,500 m)

SIZE Length: 2⅜–3⅛ in (6–8 cm). Weight: 3–4 g

STATUS Least Concern

Amethyst Woodstar

The Amethyst Woodstar has dark green-bronze upperparts with a white spot on each flank, and mottled green-gray underparts. The male has a brilliant, shining amethyst-colored throat above a white collar. There is a small white spot behind each eye, and the short, straight bill is black. The tail is forked and the dark purple-black tail feathers have white tips. The female's throat is white, mottled with green, and she has a white line running from behind the eye down to the white collar. Her underparts are pale orange, and her dark green outer tail feathers are shorter than the male's and have a black band and pale tips. The species feeds at low to mid-levels on the nectar of a wide range of different plants and trees, and also hawks for small insects. It builds a small cup-like nest of soft plant material on a branch inside a bush or tree for protection. The total population is unknown, although numbers appear to be decreasing and the species is regarded as uncommon throughout its range.

DISTRIBUTION Eastern Colombia, southern Venezuela, the Guianas, eastern Ecuador, northeastern Peru, northern Bolivia, Brazil, Paraguay, northeastern Argentina

HABITAT Forest edges, open areas, clearings, semi-open areas, savanna, scrubland; 0–4,900 ft (0–1,500 m)

SIZE Length: 2⅜–2¾ in (6–7 cm). Weight: 2–3 g

STATUS Least Concern

MALE

Lucifer Hummingbird

The northernmost representative of the woodstar–sheartail group, the striking Lucifer Hummingbird is found in arid interior habitats from central Mexico to the southwestern United States. Both sexes have long, curved bills that are not obviously adapted to any feeding specialization. They pollinate a variety of desert plants, including desert honeysuckles (*Anisacanthus*), coralbeans (*Erythrina*), paintbrushes (*Castilleja*), and Mexican Buckeye (*Ungnadia speciosa*), but also steal nectar from large bat-pollinated agaves (*Agave*) without coming into contact with the plants' anthers or pistil. Males do not hold breeding territories, but instead they seek out and court females with a spectacular zigzagging dive display at the nests that the females have constructed. Their seasonal movements need more investigation and are not well understood, but involve both north–south and elevational migration. In the non-breeding season, Lucifer Hummingbird overlaps in range with its sister species, the nearly identical Beautiful Hummingbird.

DISTRIBUTION Mexico, United States

HABITAT Arroyos, thorn scrub, dry oak and pine–oak woodland in rocky canyons, tropical deciduous forest in non-breeding season; 3,950–7,400 ft (1,200–2,250 m)

SIZE Length: 3½–3⅞ in (9–10 cm). Weight: 3–4 g

STATUS Least Concern

FEMALE

Black-chinned Hummingbird

A common and adaptable species from western parts of North America, the Black-chinned Hummingbird prospers in a variety of natural and altered habitats. It is a supreme generalist, equally at home in desert valleys, mountain woodlands, lush streamside forests, suburban gardens, and urban parks. The shapes of the inner primary feathers of adult males create a soft but distinctive whine in flight. Females and young males are similar in appearance to those of the Ruby-throated Hummingbird, but with drabber plumage and longer bills. Black-chinned Hummingbirds are relatively timid and tend to be dominated by other species, especially Rufous, Anna's, and Ruby-throated Hummingbirds. Most of the breeding population retreats southwards to Mexico in the winter, though increasing numbers are wintering around gardens and feeders in the southeastern United States.

MALE

DISTRIBUTION Breeds from south-central and western United States and extreme southwestern Canada to north-central Mexico; winters in central and southwestern Mexico

HABITAT A wide range of woodland types, forest edges, orchards, urban parks, gardens; tropical deciduous forest and thorn scrub in winter; 0–6,900 ft (0–2,100 m)

SIZE Length: 3⅜–3¾ in (8.5–9.5 cm). Weight: 2.5–4.5 g

STATUS Least Concern

Ruby-throated Hummingbird

The Ruby-throated Hummingbird is the only member of its family to nest in the eastern United States and Canada. Though not as comfortable in suburbia as Black-chinned and Anna's Hummingbirds, the species is still a familiar sight at garden flowers and feeders, especially in late summer. In spring migration, much of the eastern population joins millions of northbound songbirds in crossing the Gulf of Mexico, a journey of 500 miles (800 km) over open ocean. To accomplish this feat, they must store enough fat to double their body weight. On arrival in the United States, they refuel on the nectar of spring flowers such as Spotted Jewelweed (*Impatiens capensis*) and tree sap from wells drilled by sapsuckers. In fall migration, most Ruby-throateds seem to take the overland route around the Gulf, often congregating by the thousands in coastal Texas.

DISTRIBUTION Breeds in southeastern and south-central Canada, eastern United States; winters from southeastern United States south to Costa Rica

HABITAT Broadleaf, conifer, and mixed woodland and forest, forest clearings and edges, orchards, parks, gardens; secondary growth, savannas, edges, orchards, pastures, and other open habitats in winter; 0–9,850 ft (0–3,000 m)

SIZE Length: 3⅜–3¾ in (8.5–9.5 cm). Weight: 2.5–5.5 g

STATUS Least Concern

MALE

MALE

Mellisuga minima

Vervain Hummingbird

This diminutive hummingbird is proportionately large-headed, and has a short bill and very short tail. The plumage is shining green above and white below with a speckled throat, and the male also has green flanks. The male's tail is notched, while the female's is wedge-shaped with white edges. Subsp. *minima* is lighter in color than subsp. *vielloti*. The Vervain Hummingbird has been observed to take nectar from a very wide range of flower types, including tiny species usually pollinated by insects, and it has a very rapid feeding rate. The male's notably loud twittering song is given from a high, exposed perch—typically, each male has five or so favorite song perches within its territory. Approaching females are treated to a circling display flight. The female builds a tiny nest in which the eggs are incubated for 12 days. This is a common though still rather little-known species.

DISTRIBUTION Subsp. *minima* is found across Jamaica; subsp. *vielloti* occurs on Hispaniola and other small islands nearby

HABITAT Open woodland, parks, gardens, scrub; 0–6,550 ft (0–2,000 m)

SIZE Length: 2⅜–2¾ in (6–7 cm). Weight: 2.5–3.5 g

STATUS Least Concern

Bee Hummingbird

The smallest bird in the world, this short-tailed and short-billed species is also strikingly colorful. Both sexes are blue–green above and whitish below, but the male's head and elongated neck feathers are metallic pinkish red. This species visits mainly small flowers, and like other very small species has an extremely high metabolic rate, so must feed rapidly and often. It is an avid flycatcher, and even chicks still in the nest will catch passing insects. The species is classed as Near Threatened because of a rapid decline in its numbers. It has become scarce in Cuba and has disappeared from the nearby Isla de la Juventud, the only other place it occurred, probably due to the loss of mature woodland on the island. To secure the future of this unique and remarkable bird, habitat destruction must be curtailed and better protection afforded to the nature reserves where it occurs.

DISTRIBUTION Cuba

HABITAT Woodland, scrub, established gardens, marshes; 0–5,900 ft (0–1,800 m)

SIZE Length: 2–2⅜ in (5–6 cm). Weight: 1.5–2 g

STATUS Near Threatened

MALE

MALE

Calypte anna

Anna's Hummingbird

Few hummingbirds are as comfortable in human environments as Anna's Hummingbird. Since the 1930s, it has expanded its breeding range from coastal portions of California and northern Baja California north to southwestern British Columbia and inland to urban oases in the deserts of the southwestern United States. It takes advantage of a variety of non-native nectar sources, from the South American Tree Tobacco (*Nicotiana glauca*) to African aloes (*Aloe*) and Australian gum trees (*Eucalyptus*). Its abundance in urban settings has made Anna's Hummingbird a convenient subject for research, and as a result it is one of the most thoroughly studied members of its family. Ultra-high-speed video has revealed that the male's dive noise, produced at the bottom of the arc during the elliptical dive display, is created by vibration of the outer tail feathers as air rushes over them at 60 mph (97 kph).

DISTRIBUTION Breeds from extreme northwestern Mexico (Baja California Norte) along the Pacific Coast to extreme southwestern Canada and inland to western Texas; many move north, south, or inland after breeding; range increasing

HABITAT Year-round in coastal scrub and chaparral, riparian and canyon woodlands, residential areas and parks; mountain meadows and forest edges in late summer; 0–5,900 ft (0–1,800 m) in the breeding season, to over 10,000 ft (3,050 m) post-breeding

SIZE Length: 3½–3⅞ in (9–10 cm). Weight: 3–5.5 g

STATUS Least Concern

FEMALE

Costa's Hummingbird

With its small size, Costa's Hummingbird is well adapted to the intense heat and scarce resources of its desert home. The species is partially migratory, expanding northward in late winter and spring, and southward into Mexico during the fall and early winter. The relatively drab females build nests in shaded sites low in small trees, shrubs, vines, and cacti. Soft plant materials used in nest construction help to insulate the eggs and young from temperature extremes. During the late nesting season, when temperatures may exceed 100°F (38°C), the birds spend the middle of the day in the shade of desert vegetation, emerging near sunset to take nectar before dusk. Nectar sources include many insect-pollinated plants such as Desert Lavender (*Hyptis emoryi*) and Creosotebush (*Larrea tridentata*), as well as more typical hummingbird flowers such as Chuparosa (*Justicia californica*) and Ocotillo (*Fouquieria splendens*).

DISTRIBUTION Southwestern United States and northwestern Mexico

HABITAT Desert scrub, chaparral, riparian thickets, and fringes of desert cities year-round; foothills, mountain canyons, and tropical thorn scrub in the non-breeding season; 0–6,900 ft (0–2,100 m)

SIZE Length: 3–3½ in (7.5–9 cm). Weight: 3–3.5 g

STATUS Least Concern

MALE

Bumblebee Hummingbird

The Bumblebee Hummingbird earns its name as much for its insect-like style of flight as for its small size. All plumages are superficially similar to those of the slightly larger Calliope Hummingbird, which overlaps the Bumblebee's range in migration and in winter. Adult males have solid fuchsia-pink gorgets and white-tipped outer tail feathers more typical of females and young males in most of its close relatives. Subsp. *margarethae* males are slightly smaller and have a darker, amethyst-violet throat patch. The male's song consists of shrill, siren-like whistles punctuated by clicks. Males are noisy in flight as well, with narrowed outer primary feathers that create a low nasal buzz like the drone of a bumblebee. The species feeds on invertebrates and a variety of flowers, often by sneaking into the territories of other hummingbirds. Foraging near the ground helps them to avoid detection by larger aggressors.

FEMALE

MALE

DISTRIBUTION Subsp. *heloisa* occurs in northeastern, central, and southern Mexico; subsp. *margarethae* occurs in western and northwestern Mexico

HABITAT Edges, clearings, and secondary growth in humid to semihumid conifer and pine–oak forest and woodland; 4,900–9,850 ft (1,500–3,000 m)

SIZE Length: 2¾–3 in (7–7.5 cm). Weight: 2–3 g

STATUS Least Concern

Wine-throated Hummingbird

The Wine-throated Hummingbird is the
southern counterpart of the Bumblebee
Hummingbird, replacing it south of the
Isthmus of Tehuantepec. Though the two
are so similar in appearance that at times they
have been considered one species, they do not
overlap in range and differ in several significant
ways. Male Wine-throateds sing far more
varied songs than male Bumblebees, the song
consisting of a rapid staccato series of squeals,
chips, and twitters, and they lack narrowed tips
on the outer primary feathers. The tips of the
outer tail feathers are white in adult males of
Mexican and Guatemalan populations, and
strongly suffused with cinnamon in the
Honduran subsp. *selasphoroides*. They share
feeding strategies with the Bumblebee
Hummingbird, using their small size and insect-
like flight to escape detection while foraging
within the territories of larger hummingbirds.

MALE

DISTRIBUTION Subsp. *ellioti*
occurs in southeastern Mexico
and southern Guatemala;
subsp. *selasphoroides* is found
in western Honduras

HABITAT Edges, clearings, and
secondary growth in humid to
semihumid conifer and pine–
oak forest and woodland;
4,900–9,850 ft (1,500–3,000 m)

SIZE Length: 2½–2¾ in
(6.5–7 cm). Weight: 2–3 g

STATUS Least Concern

Broad-tailed Hummingbird

The silvery wing trill of the male Broad-tailed Hummingbird, created by its modified outer-wing feathers, is an iconic sound of spring and summer in the southern and central Rocky Mountains. Females and young males are comparatively drab except for their tails, which are banded in rufous, green, black, and white. Annual and perennial flowers such as Skyrocket (*Ipomopsis aggregata*), larkspurs (*Delphinium*), red columbines (*Aquilegia*), Glacier Lily (*Erythronium grandiflorum*), and paintbrushes (*Castilleja*) are important nectar sources in summer, but migrants that arrive before the first flowers bloom will take advantage of sweet, nutritious willow sap from wells drilled by sapsuckers. They also hunt flying insects and glean small invertebrates such as aphids and spiders from flowers and foliage. A banded female Broad-tailed Hummingbird that reached a minimum age of 12 years holds the longevity record for North American hummingbirds.

DISTRIBUTION Rocky Mountains of northwestern United States south through Mexico to Guatemala; northern populations winter in Mexico, occasionally in lower elevations in the southern United States

HABITAT Conifer, broadleaf, and mixed forests, clearings, and meadows in mountains, foothills and intermountain valleys; 2,300–10,500 ft (700–3,200 m). Northern migrants winter in conifer and mixed forests, dry thorn forest, and gardens

SIZE Length: 3½–3⅞ in (9–10 cm). Weight: 3–4.5 g

STATUS Least Concern

MALE

Calliope Hummingbird

Formerly the sole member of the genus *Stellula* ("little star"), the diminutive Calliope Hummingbird is the smallest breeding bird in the United States and Canada and one of the smallest birds in the world. Though often pugnacious for their size, Calliopes are usually subordinate to larger species, especially the savagely territorial Rufous Hummingbird. When competitors exclude them from typical hummingbird-pollinated flowers, their short, thin bills allow them to forage at insect-pollinated flowers. During courtship, the male performs a spectacular "shuttle display," weaving to and fro in front of the female with the "rays" of his gorget extended in a wine-purple starburst. For protection from the cold mountain climate, females site their nests under overhanging branches.

MALE

FEMALE

DISTRIBUTION Breeds in western United States, southwestern Canada, and extreme northern Baja California; migrates mainly through the Rocky Mountains and Sierra Madre Occidental; winters in southwestern Mexico and occasionally at low elevations in the southern United States

HABITAT Aspen and willow thickets, stream sides, open conifer forests, young secondary growth following wildfires or logging; as low as 600 ft (180 m) but mainly 3,950–11,150 ft (1,200–3,400 m)

SIZE Length: 3–3⅛ in (7.5–8 cm). Weight: 3–3.5 g

STATUS Least Concern

Rufous Hummingbird

The notorious pugnacity of the Rufous Hummingbird helps to ensure it has a steady supply of nectar to fuel its epic migrations. A banded female set the record for the longest documented migration of any hummingbird: a minimum of 3,500 miles (5,600 km) between Alaska and Florida, although the species normally winters in Mexico. The backs of adult males may be entirely fiery rufous or mixed with varying amounts of iridescent green. Females and young males are predominantly bright green, with tails boldly banded in rufous, black, and white. Sharp tips on the outer primary feathers of adult males create a sinister whine in flight, and young and old birds alike screech and chatter in combat over choice nectar sources. The species' winter range has expanded north and east in recent years, assisted by gardens with feeders and winter-blooming exotic plants, but the National Audubon Society estimates a 60 percent population decline since 1967.

DISTRIBUTION Western Canada, United States, Mexico

HABITAT Nests in mature mountain conifer, broadleaf, and mixed forests, forest edges, clearings, secondary growth, parks, and gardens; 0–6,000 ft (0–1,850 m). Migrates and winters in varied habitats including alpine meadows, riparian forest, oak woodlands, desert scrub, tropical deciduous forest, and gardens

SIZE Length: 3½–3⅞ in (9–10 cm). Weight: 3–4.5 g

STATUS Least Concern

MALE

Allen's Hummingbird

A relatively mild-mannered southern counterpart of the Rufous Hummingbird, Allen's Hummingbird breeds along the southern and central Pacific coast of the United States. There are two distinct populations: a northern migratory subspecies, subsp. *sasin*, and a nonmigratory population originally from the Channel Islands off California, subsp. *sedentarius*, that colonized the Los Angeles area in the 1960s and has spread north and south along the coast. The mild climate allows nonmigratory females to nest up to four times per year. Allen's Hummingbirds are important pollinators of several of California's endemic and near-endemic plants, including Orange Bush Monkeyflower (*Mimulus auruntiacus*), Scarlet Larkspur (*Delphinium cardinale*), Showy Island Snapdragon (*Galvezia speciosa*), and the endangered Western Lily (*Lilium occidentale*).

DISTRIBUTION Subsp. *sasin* ranges from coastal extreme southwestern Oregon to southern California in the breeding season, and winters in south-central Mexico, rarely reaching east to Texas and the southeastern United States; subsp. *sedentarius* occurs in coastal southern California and on the Channel Islands (Anacapa, Santa Catalina, San Clemente, Santa Cruz, San Miguel, Santa Rosa) year-round

HABITAT Coastal scrub and chaparral, mixed evergreen forest, riparian woodland, parks, gardens; 0–1,200 m (0–3,950 ft)

SIZE Length: 3⅛–3½ in (8–9 cm). Weight: 3–3.5 g

STATUS Least Concern

MALE

Volcano Hummingbird

This is a dainty, short-billed·hummingbird with the typical woodstar pattern, the male being green with a white breast band and flared, glittering throat patch, and the female similar but with a dark-speckled whitish throat. The color of the male's throat patch varies between the three subspecies: Subsp. *flammula* has a pinkish-purple throat patch; subsp. *simoni* has a pinkish-red throat patch; and subsp. *torridus* has a purplish-gray or purplish-green throat patch. The species visits a range of mainly small flowers, including *Salvia*, *Digitalis*, and *Rubus*, and also accesses nectar in longer holes via holes made in their bases by bumblebees and flowerpiercers. Males establish and defend a territory when courting, and perform various aerial displays to attract females. The female's nest is a tiny cup made from cobwebs, moss, lichen, and plant down. As a species of open habitats, this bird may have benefitted from deforestation in its range.

FEMALE

MALE

DISTRIBUTION Subsp. *flammula* occurs on the volcanoes Irazú and Turrialba in central Costa Rica; subsp. *simoni* occurs on the volcanoes Poás and Barva in central Costa Rica; and subsp. *torridus* is found in south Costa Rica and west Panama

HABITAT Quite open habitats on volcanic slopes, including scrubland, thick grassland, paramo, and open bogland; 5,900–11,500 ft (1,800–3,500 m)

SIZE Length: 3–3⅛ in (7.5–8 cm). Weight: 2.5–3 g

STATUS Least Concern

Scintillant Hummingbird

This is one of the smallest hummingbird species, and has compact proportions. Both of the sexes are shining green above, though the female is more olive-toned, and both have pale buff-tinged underparts. The male has a broad pinkish-orange throat patch, with flared and elongated feathers at the sides, while the female's throat is white with fine, dark speckles. The species is a trapliner and a discreet raider of larger hummingbirds' territories—many of the flowers it visits are tiny and more usually pollinated by insects. When flycatching it usually hawks from a perch, and it also picks tiny invertebrates from foliage. Males hold territories for courtship purposes, proclaiming ownership with a diving aerial display. The nest is a tiny cup of pale, soft plant down, often built in long grass or low in a shrub. This bird's population is stable.

FEMALE

MALE

DISTRIBUTION North-central Costa Rica to west Panama, mainly on the Pacific side

HABITAT Thickets and scrubby grassland, forest edges, low secondary growth, gardens, plantations; 2,950–6,900 ft (900–2,100 m)

SIZE Length: 2½–2¾ in (6.5–7 cm). Weight: 2–2.5 g

STATUS Least Concern

Hook-billed Hermit

This hermit has a rather straight bill with a small hook at the tip. The plumage is warm olive green above and rich, bright cinnamon below. The white-tipped tail is wedge-shaped without elongated central feathers. The bird feeds mainly in the forest understory, favoring *Heliconia* flowers, and like other hermits it is a trapliner. The nest is a typical hermit construction at the tip of a hanging leaf, with incubation taking about 15 days and fledging between 20 and 27 days. This bird was considered Critically Endangered up until 2000, when surveys revealed it was more common than had been supposed. However, its estimated population is still very small (probably between 350 and 1,500 individuals) and fragmented, because of widescale habitat loss in its small range. It occurs in at least two nature reserves but these are under pressure and require more effective protection.

DISTRIBUTION Southeast Brazil

HABITAT Damp forest; 0–1,650 ft (0–500 m)

SIZE Length: 4¾–5⅛ in (12–13 cm). Weight: 5.5–7 g

STATUS Endangered

Pale-tailed Barbthroat

Though not common anywhere, the Pale-tailed Barbthroat is found over much of Amazonia. The species is divided into four subspecies that differ slightly in color and pattern. The belly and facial markings vary in contrast, and the outer tail feathers range from white in the nominate subsp. *leucurus* ("white-tailed") to pale ocher-buff in the northwestern subsp. *cervinicauda* ("deer-tailed"). Males deliver their high-pitched warbling songs and perform tail-bobbing displays on thin horizontal twigs in the forest understory. When another barbthroat visits the singing territory, the male will hover in front of the visitor with his tail fanned and throat pattern displayed while singing a muted version of his song, typically followed by a chase through the forest. Young males often loiter on the fringes of singing territories and may briefly take over song perches and practice their own displays while the owner is absent.

DISTRIBUTION Subsp. *leucurus* ranges from Venezuela to Suriname, Amazonian Brazil, and northern Bolivia; subsp. *cervinicauda* occurs in eastern Colombia, eastern Ecuador, northeastern Peru, and western Amazonian Brazil; subsp. *medianus* occurs in northeastern Brazil south of the Amazon; subsp. *rufigastra* ranges from central Peru to northern Bolivia

HABITAT Open understory of humid lowland and mountain forests, riversides, swamps, secondary growth, plantations; 0–5,250 ft (0–1,600 m)

SIZE Length: 3⅞–4⅜ in (10–11 cm). Weight: 5–7 g

STATUS Least Concern

Sooty Barbthroat

The Sooty Barbthroat is considered by some simply to be a melanistic form of the Pale-tailed Barbthroat, and the exact relationship and distribution of the two species is not well understood. Aside from the nominate there is one subspecies, *loehkeni*, found in the same region as subsp. *niger* but paler in color; the two are sometimes treated as separate species. Both subspecies have dark green tails that lack the distinctive pale outer tail feathers of the Pale-tailed Barbthroat. The Sooty Barbthroat is uncommon and local within its relatively small range, making it more vulnerable to the ongoing deforestation of the lower Amazon. Like other hermits, it forages by traplining between patches of *Heliconia* and other flowers in the forest understory, and hover-gleaning insects and spiders from foliage.

DISTRIBUTION Subsp. *niger* occurs in French Guiana and adjacent Brazil (northern Amapá); subsp. *loehkeni* is found in northeastern Brazil, north of the Amazon

HABITAT Understory of lowland evergreen forest; 0–5,250 ft (0–1,600 m)

SIZE Length: 3⅞–4⅜ in (10–11 cm). Weight: 5–7 g

STATUS Least Concern

Broad-tipped Hermit

The Broad-tipped Hermit is a fairly small, slenderly built bronzy-brown hummingbird, with a rather full rounded tail and a long, gently downcurved, dark-tipped bill that has an unusually wide base. The distinctive bill and tail shape are the reasons for its classification in a separate genus to the otherwise similar *Phaethornis* hermits. The wide white tips to its tail feathers are striking when the bird is in flight. The sexes look the same. It is rather uncommon and very little known. It inhabits dense, thorny caatinga forest, and like other hermits is a trapliner, working its way around a predictable circuit from flower to flower when feeding. Breeding females in particular also take many small insects, including flies plucked from spiders' webs. Two eggs are laid and the incubation period is 14 days, the chicks fledging at about 20 days old.

DISTRIBUTION Eastern Brazil

HABITAT Humid, dense caatinga thorn forest; 1,650–2,300 ft (500–700 m)

SIZE Length: 4⅜ in (11 cm). Weight: 3 g

STATUS Least Concern

Tapajos Hermit

This is a tiny, mainly dark gray-green hummingbird, with prominent white eyebrows and an orange belly and rump. It has had a confused history in terms of classification. It was first described in 1950 as a subspecies of the Little Hermit, but reappraised in 1996 as a population of hybrids between Streak-throated and Reddish hermits. Field studies on its plumage and distribution, however, refuted this idea—for example, the researchers found that the Tapajos Hermit has some consistent plumage features not found in either the Streak-throated or Reddish Hermits. In 2009 it was officially reclassified as a full "new" species. Found in only a small area of forest, where logging and development are rife, it is thought to be rare and declining, although detailed population studies have yet to be carried out. It is known to occur in some protected areas.

DISTRIBUTION Recorded only around the Teles Pires, Tapajós, and Xingu rivers in Brazil

HABITAT Mainly primary rainforest, but has been observed in secondary growth; 0–1,650 ft (0–500 m)

SIZE Length: 3½ in (9 cm). Weight: 2.5–4 g

STATUS Near Threatened

Cinnamon-throated Hermit

A fairly small hermit, this hummingbird has strong, warm orange tones on its underside, tail edges, and rump, with the male's coloration generally darker than the female's, especially on the throat. The white tail tip is more elongated and sharply pointed than in most similar-sized hermits. There are two populations, one in northeast Brazil and the other in eastern Bolivia and southwest Brazil, but it is likely that this species is also found between these two areas. The species makes use of a range of habitat types from caatinga to riverine woodland, but avoids dense primary rainforest. On its foraging circuits it visits such flowers as *Pavonia*, *Ruellia*, and *Bauhinia*. The male's song, given at the lek, alternates very high, thin notes with much lower-pitched phrases. The nest takes the typical hermit form, a pouch fixed to a hanging leaf tip, but little is known of its breeding biology.

DISTRIBUTION Northeast Brazil, southwest Brazil, and eastern Bolivia, and possibly continuously between these areas

HABITAT Various well-vegetated habitats; 0–1,650 ft (0–500 m)

SIZE Length: 3⅞ in (10 cm). Weight: 2.5–3 g

STATUS Least Concern

Black-throated Hermit

Reddish Hermit

Similar to the Reddish Hermit, this is an attractive, very small hermit, that has a blackish central patch on the breast (sometimes extending to form a complete band), separating the orange belly from the dark-streaked throat. Its whitish undertail-coverts are diagnostic. The sexes are slightly different, the male having a darker throat, more prominent dark breast marking, and, usually, shorter wings than the female. There are two subspecies, subsp. *atrimentalis* and the larger subsp. *riojae*. It is not a well-studied species, though is likely to have typical hermit feeding and breeding behavior. The flower types it is known to visit include *Aechmea*, *Costus*, and *Drymonia*, and it will also opportunistically take very small flies and other insects. Although the species is currently not of conservation concern, it is declining due to habitat loss and a greater understanding of its biology will be needed for future conservation. It does occur in protected areas in Colombia.

The Reddish Hermit is one of the smallest hummingbirds. Both the male and female are dark green above and pale orange-red on their lower backs and underparts, with a dark mask, white stripes above and below the eye, and a downcurved bill. The male often has a dark patch on his breast, which can form a band, while the female has a usually lighter breast and a longer tail and wings. The male of subsp. *episcopus* tends to be more orange underneath; subsp. *nigricinctus* is smaller, with more reddish underparts and white tail margins; and subsp. *longipennis* is larger, with a white chin and reddish margins and tips to the tail feathers. The species is a trapliner, feeding on nectar; it also eats small spiders and insects. The cone-like nest is made from plant fibers, leaves, lichen, and moss, and is connected to an overhanging branch or leaf, usually 3 to 6 feet (1 to 2 m) above ground. The species is common throughout its range, but its population trend appears to be downward.

DISTRIBUTION Subsp. *atrimentalis* occurs in the Andes of east Colombia, Ecuador, and north Peru; subsp. *riojae* occurs in central Peru

HABITAT Rainforest edges, secondary growth; usually 0–1,650 ft (0–500 m) but up to 3,950 ft (1,200 m)

SIZE Length: 3⅛–3½ in (8–9 cm). Weight: data not available but probably 2–3.5 g

STATUS Least Concern

DISTRIBUTION Subsp. *ruber* ranges from Suriname to French Guiana, Brazil, southeastern Peru, and northern Bolivia; subsp. *episcopus* occurs in central and eastern Venezuela, Guyana, and northern Brazil; subsp. *nigricinctus* occurs in far southwestern Venezuela, eastern and southern Colombia, and northern Peru; subsp. *longipennis* occurs in southern Peru

HABITAT Rainforest, forest edges, savanna woodland, scrubland, thickets; 0–4,600 ft (0–1,400 m)

SIZE Length: 3–3½ in (7.5–9 cm). Weight: 2–3 g

STATUS Least Concern

White-browed Hermit

This small hermit is very similar to the Reddish Hermit and is sometimes considered to be a subspecies of that bird. However, the two species have adjoining distributions and only one case of natural hybridization between them is known, which suggests they are in fact distinct species. The White-browed Hermit has a white tail tip, while that of the Reddish Hermit is orange. Both sexes have orange undersides and greenish-brown upperparts; females are paler than males. This is a little-known species of terra firma (unflooded) rainforest; it will also use scrubland, bamboo thickets, and other well-vegetated habitats. Its habits and breeding behavior are probably very similar to those of its close relatives, but further ecological study is needed—especially as its known range is quite restricted, its distribution within that range is patchy, and its population is thought to be decreasing.

DISTRIBUTION A narrow strip from southeast Peru to central Bolivia

HABITAT Fairly dry forest and scrubland; 0–3,300 ft (0–1,000 m)

SIZE Length: 3½ in (9 cm). Weight: 2–3 g

STATUS Least Concern

Buff-bellied Hermit

This is a fairly small hermit, with a long, almost straight bill and distinctly elongated central tail feathers—these and the other tail feathers have broad white tips. Otherwise, the bird is mainly rather dull gray-brown with a little green iridescence on the back and a pale orange underside. The sexes are similar, while juveniles have pale feather fringes, giving them a slightly scaly appearance. The species prefers drier woodland and forest habitats, and like other hermits is a trapline feeder, visiting successive flowers along a regular route. This strategy gives the flowers enough time to replenish their nectar stocks before the hummingbird's next visit. Although this is quite a commonly encountered species, its habits are not yet well known. The nest takes the typical conical hermit form, and the eggs hatch after 14 to 15 days of incubation; the young fledge 20 days later.

DISTRIBUTION East of the Andes in Bolivia and into west Brazil

HABITAT Dry forest and scrubland on hilly ground; 500–2,600 ft (150–800 m)

SIZE Length: 4⅜–4¾ in (11–12 cm). Weight: 3.5–4 g

STATUS Least Concern

Pale-bellied Hermit

This is a smartly marked hermit, with dark, glossy green upperparts, shading to bronze on the rump, contrasting with a whitish underside. The female's bill is more curved than the male's, and her wings are shorter. There are two subspecies: subsp. *anthophilus*; and the slightly paler, more blue-toned subsp. *hyalinus*, found only on islands in the Gulf of Panama. Like other hermits, the Pale-bellied forages at low levels in the understory, moving along a favorite route and stopping at flowers such as *Heliconia* and *Brownea*. It shows a preference for drier woodlands more than most other hermit species and will also use degraded forests. Males display communally at leks, and females select a partner on the basis of song and display quality. There is some evidence that the species may undertake short seasonal migrations. It is quite common in some areas, in particular coastal Venezuela.

DISTRIBUTION Subsp. *anthophilus* ranges from central Panama into Colombia and north Venezuela; subsp. *hyalinus* occurs only on the Pearl Islands in the Gulf of Panama

HABITAT Semi-dry woodland and scrub; 0–4,900 ft (0–1,500 m)

SIZE Length: 5⅛ in (13 cm). Weight: 4–5.5 g

STATUS Least Concern

Needle-billed Hermit

A medium-sized hermit, this bird has a straighter bill than most other *Phaethornis* species, and has a rich orange throat, breast, and belly. Its bronze-green upperparts are also strongly suffused with rich orange tones. The female is slightly smaller than the male, and has a shorter bill and wings, but otherwise has a similar plumage. This species is very similar to the rare Koepcke's Hermit, but it lacks that species' dark stripe on the side of the throat and is generally found at lower altitudes, where it may be quite common, particularly in the western part of its range. It feeds mainly at low levels and, like other hermits, forages by traplining, checking feeding sites along a route that may be 0.5 miles (800 m) long or more. Unlike other hermits it is not known to lek. It builds the typical hermit-style pouch nest on a hanging leaf.

DISTRIBUTION Eastern Peru, north Bolivia and western Brazil

HABITAT Lowland rainforest, using the understory, also plantations and bamboo thickets; 0–1,050 ft (0–350 m)

SIZE Length: 4¾ in (12 cm). Weight: 4–6 g

STATUS Least Concern

Hyacinth Visorbearer

This is a small greeny-bronze hummingbird with a short tail. The male is dark blue underneath, with a white spot behind the eye, and a bright green-gold throat patch, blue throat band, and white breast band. The female is similar but has speckled gray-blue underparts. Subsp. *ilseae* has a deep violet neck and rich violet-blue underparts, while subsp. *soaresi* is slightly larger and has a blue line separating the violet and black on its neck. The bird darts to and from its perch to feed on insects, and takes nectar from flowers at about head height, favoring bromeliads, cacti, Verbenaceae, and Loranthaceae. It builds a small cup-like nest from seeds and leaves in forked gaps between branches. The bird is common locally, although the population is thought to have been larger in the past. Numbers have not been quantified, but the species is considered stable, although habitat loss is a concern.

DISTRIBUTION All subspecies can be found in eastern-central Brazil: subsp. *scutatus* occurs at high elevations in central and eastern Minas Gerais; subsp. *ilseae* occurs at moderate heights in central and eastern Minas Gerais; subsp. *soaresi* occurs in south-central Minas Gerais

HABITAT Mostly mountainous, arid, and rocky areas at higher elevations, also valleys with dense forest; 2,950–5,900 ft (900–1,800 m)

SIZE Length: 3⅛–3¾ in (8–9.5 cm). Weight: 3.5–5 g

STATUS Near Threatened

Black-eared Fairy

This very widespread hummingbird resembles the Purple-crowned Fairy and replaces it east of the Andes, but the male has a green rather than violet crown, and there is less difference in tail length between the sexes. It has three subspecies—subsp. *auritus*, subsp. *phainolaemus*, and subsp. *auriculatus*—the males of which vary in the extent of green on the throat and chin. It is predominantly a lowland species, and feeds mainly at higher levels in the forest, taking nectar directly and by flower-piercing. The nest, a neat cup of plant-seed down bound together with spider webs, is built on a vertical branch well above the ground—sometimes as high as 100 ft (30 m). The incubation period is 15 to 16 days, and chicks fledge at between 23 and 26 days. Although the species has a very extensive range, it is rather uncommon in most areas and is declining.

DISTRIBUTION Subsp. *auritus* occurs in southeast Colombia and east Ecuador through north Brazil to northeast Venezuela and the Guianas; subsp. *phainolaemus* occurs in north-central Brazil; subsp. *auriculatus* occurs from east Peru and central Bolivia to central-eastern Brazil

HABITAT Wet lowland forest and secondary growth; 0–2,600 ft (0–800 m) but usually below 1,300 ft (400 m)

SIZE Length: 3⅞–5⅜ in (10–13.5 cm). Weight: 5.5–6.5 g

STATUS Least Concern

White-tailed Goldenthroat

This is a small, large-headed, short-tailed hummingbird, the male olive green above and brighter yellow-green below with white tail sides, and the female drabber and paler with a white tail tip. There are three subspecies: subsp. *guainumbi*; subsp. *andinus*, which has more white in the tail; and subsp. *thaumantias*, which is shorter-billed, more orange-toned, and has less white in the tail. The bird forages very low through the vegetation, taking nectar from all kinds of flowers and gleaning spiders and insects from foliage—some insects are also caught on the wing. It builds a soft cup nest from bulrush down camouflaged with lichens, usually low in a bush and often overhanging water. A common species in much of its range, it will visit gardens and in good habitat several females may nest close together, though it is not social when feeding.

DISTRIBUTION Subsp. *guainumbi* occurs in the Guianas, north Brazil, Venezuela, and Trinidad; subsp. *andinus* occurs in eastern Colombia; subsp. *thaumantias* ranges through eastern Bolivia and Paraguay down through central Brazil to northeast Argentina

HABITAT Grassland, swamps, scrubland, gardens and other open habitats; 0–1,950 ft (0–600 m)

SIZE Length: 3¾–3⅞ in (9.5–10 cm). Weight: 4.5–5 g

STATUS Least Concern

Tepui Goldenthroat

The shining green plumage of this large hummingbird is interrupted only by a white tail base and tail tip, blackish wings, and dark speckles on the underside in the female. It is the only goldenthroat lacking white facial markings. As is typical of the group, it feeds at very low levels, and is mainly a trapline feeder moving from flower to flower and also hawking insects. It works along a regular route and does not tend to linger long at any one spot. This habit, along with its reluctance to perch in clear view and its liking for thick scrubland habitats, makes it a difficult bird to observe. The cone-shaped nest, however, is sometimes placed in quite exposed situations, though its covering of leaf fragments and lichen helps disguise it. The species undertakes altitudinal movements outside the breeding season, and within its limited range is fairly common.

DISTRIBUTION Mountainous parts of Venezuela and north Brazil

HABITAT Scrubland and cloud forest edges; 4,250–7,200 ft (1,300–2,200 m)

SIZE Length: 4⅜–4¾ in (11–12 cm). Weight: 4–6 g

STATUS Least Concern

Fiery-tailed Awlbill

This unusual small hummingbird is related to the mangoes (genus *Anthracothorax*) and is sometimes classed with them. It has a short, straight bill with a pronounced upwards curve at the tip. The male is glossy green with a blackish belly and violet-gold tail sides. The female has white underparts with a central black belly stripe. This is a low-level feeder and a trapliner, visiting *Clusia* and *Dioclea* shrubs and taking many insects, both in flight and by gleaning from the undersides of leaves. The nest is built quite high in a tree or large shrub, and is a tiny cup fixed to a broad horizontal branch. This bird was formerly considered to be Near Threatened but was reassessed as Least Concern in 2004. Although it is probably declining and seems rare in many areas, it remains very widely distributed.

DISTRIBUTION Southeast Venezuela, the Guianas, and north-central Brazil, plus a small separate population in east Ecuador

HABITAT Large forest clearings with rocky outcrops and savanna-like vegetation; 0–1,650 ft (0–500 m)

SIZE Length: 3⅛–3⅞ in (8–10 cm). Weight: 4.5 g

STATUS Least Concern

Green-throated Mango

The Green-throated Mango is one of four similar members of the genus found from Mexico to South America. Adult males differ from the Black-throated Mango in that the broad black midline stripe of the underparts extends only from belly to breast, ending at a brilliant green throat. The outer tail feathers are vivid wine purple to magenta, narrowly bordered in blue-black. Adult females are boldly marked, with a velvety black and iridescent green midline stripe on the underparts bordered by snowy white, and outer tail feathers banded in wine purple, blue-black, and white. Young birds resemble adult females except for the presence of bright cinnamon coloring in the face and underparts. Declining populations in Trinidad are attributed to the destruction of mangroves.

DISTRIBUTION Trinidad, northeastern Venezuela, the Guianas, coastal northeastern Brazil and inland along the Amazon River; decreasing but still locally common over much of its range

HABITAT Open lowland habitats, marshy savannas, mangroves, secondary growth; 0–1,640 ft (0–500 m)

SIZE Length: 4⅛–4⅞ in (10.5–12.5 cm). Weight: 5.5–9 g

STATUS Least Concern

Green Mango

The Green Mango is more similar to the Green-throated Carib of eastern Puerto Rico than to other members of its genus. Both sexes and all ages are very similar in appearance, though young birds may have a brownish tinge to the upperparts and pales edges on the feathers of the underparts. Though Green Mangoes feed extensively on insects and spiders, their abundance in a given habitat depends in large part on the availability of their favorite long-tubed flowers. Important native nectar plants include the leguminous vine Bejuco de Alambre (*Neorudolphia volubilis*) and the red-flowered bromeliad *Pitcairnia bromeliifolia*, but they will also take advantage of exotic ornamental flowers such as hibiscus. They compete for nectar with the small Puerto Rican Emerald hummingbird, and the Bananaquit (*Coereba flaveola*), a songbird, both of which steal nectar by piercing the bases of flowers.

DISTRIBUTION Puerto Rico

HABITAT Mountain forests and shade coffee plantations; mainly 2,600–3,950 ft (800–1,200 m)

SIZE Length: 4⅜–5½ in (11–14 cm). Weight: 7 g

STATUS Least Concern

Merida Sunangel

The Merida Sunangel is often considered a subspecies of the Amethyst-throated Sunangel (or Longuemare's Sunangel when this is split from Amethyst-throated). The *International Union for Conservation of Nature*, which produces the Red List of Threatened Species, does not consider the Merida Sunangel a full species and so has not assessed its status. This is a small yellow-green hummingbird. The male has a light violet throat patch bordered with a fairly wide white breast band, and a small blue-green forehead patch. The female's throat is mottled green, only slightly darker than the crown, and she lacks the contrasting forehead patch. This species is territorial and aggressive, like other sunangels, and usually selects flower clumps that are at quite low levels. Within its restricted geographical range it is fairly common, though further study is needed, along with a full evaluation of its taxonomic status.

DISTRIBUTION The Andes of Mérida, northwest Venezuela

HABITAT Cloud forest edges and nearby open habitats; 6,550–11,800 (2,000–3,600 m)

SIZE Length: 3⅞ in (10 cm). Weight: 5–6 g

STATUS Not Evaluated

Bogota Sunangel

This mysterious hummingbird is known from only a single specimen, a probable male that was obtained as a skin in Bogotá in 1909. It has a dark bluish-black plumage, shading to greenish blue on the lower back and rump, and an iridescent green throat and forehead patch. The tail is deeply forked and the bill is short and straight. The correct classification of this specimen has been much disputed, with earlier authorities considering it a hybrid between the Long-tailed Sylph and the Purple-backed Thornbill, but DNA studies on tissue taken from the specimen disproved this theory and indicated the bird constituted a true species. As this bird has never been observed in the wild, it is possible only to speculate about its way of life, and indeed whether it still survives undiscovered. With no further sightings in more than a century, it is likely to be extinct or extremely rare at best.

DISTRIBUTION Unknown—probably the eastern or central Andes in Colombia

HABITAT Unknown, presumed to be humid forest at altitudes of around 8,200–9,850 ft (2,500–3,000 m)

SIZE Length: 4¾ in (12 cm). Weight: unknown

STATUS Data Deficient

Coppery Thorntail

What little is known about the Coppery Thorntail comes from two male specimens collected somewhere in South America prior to 1852. Though the labels give their origin as "Bolivia," commercially traded birds skins from the nineteenth century are notorious for vague and unreliable locality information. The specimens most closely resemble the Green Thorntail, but with coppery backs and bellies, and a band of whitish mottling across the upper breast. The pointed black outer feathers of the long, deeply forked tail have white shafts as in the Green, Black-bellied, and Wire-crested Thorntails. A pale area at the base of the bill may have been orange in life, different from any other thorntail but similar to several of the closely related coquettes. Though it hasn't been seen in over 150 years, it is possible that this enigmatic hummingbird still exists in some poorly explored corner of the Amazon Basin.

DISTRIBUTION Unknown—possibly northern Bolivia and/or adjacent Brazil

HABITAT Unknown, probably lowland rainforest

SIZE Length: male 3½ in (9 cm). Weight: unknown

STATUS Data Deficient

Racket-tailed Coquette

This species is more closely related to the coquettes (*Lophornis*) than other members of the genus *Discosura*. The male has elongated central tail-feather shafts, tipped with small kidney-shaped "rackets." Both sexes are shining olive green with a whitish-buff rump band and chestnut belly band; the male has a glittering green throat, while the female's is blackish with white edges. Its bill is short and straight. The species is a solitary canopy feeder, its hawkmoth-like appearance helping it avoid attack by more aggressive species as it sneaks into their territories. It builds a high-level nest on a horizontal branch, and the female incubates the two eggs for 13 to 14 days, before tending the chicks in the nest for 20 to 22 days. The bird has a very extensive range but is undoubtedly suffering the effects of habitat loss in many areas, as it does not accept modified habitats. It is found in some protected areas.

DISTRIBUTION Southern Venezuela, the Guianas, and northern and eastern Brazil

HABITAT Humid riverine forest and scrub; 0–650 ft (0–200 m)

SIZE Length: 3⅛–3⅞ in (8–10 cm). Weight: 3 g

STATUS Least Concern

Dot-eared Coquette

Like other coquettes, this very small hummingbird is short-billed and short-tailed, and is predominantly shining green in color with a white rump band. The male sports a neck frill of long "whiskers," which are white with green tips, and he also has a pointed orange crest. The female lacks these adornments and has an orange face. The Dot-eared Coquette is a trapline feeder, moving from flower to flower along an established circuit, and will visit a wide range of flower types. Its size and pattern give it a moth-like appearance in flight, and it has a steady, weaving flight style, less jerky than that of most other hummingbirds. Although the population size and trend of the species are not known, its conservation status was changed from Least Concern to Vulnerable in 2012 in response to the high rate of deforestation in its habitat. Its future will depend on establishing more and larger protected areas.

DISTRIBUTION North and north-central Brazil to eastern Bolivia

HABITAT Forest edges, savanna; 0–1,650 ft (0–500 m)

SIZE Length: 2¾–3 in (7–7.5 cm). Weight: 2.5–3 g

STATUS Vulnerable

Short-crested Coquette

The orange crest of the male Short-crested Coquette is a little shorter and blunter than that of the Tufted Coquette, and the black-tipped orange neck plumes are also short. In addition, the male has whitish breast and rump bands, and the throat is glittering green. The female has an orange crown and unmarked, pale buff underparts. This species feeds from flowers such as *Inga*, *Cecropia*, and *Clethra*, using all heights in the forest strata, and has typical coquette habits. Considered Endangered up until 2000, it is now classed as Critically Endangered following extensive forest clearance within its tiny range during the 1990s. Its rapid decline is continuing, and the population may now number as few as 250 individuals. The area it occupies is difficult to survey and protect, partly because of illegal drug cultivation. Without immediate establishment of an effectively protected area, this species' future is bleak.

DISTRIBUTION Southwest Mexico (Sierra Madre del Sur)

HABITAT Humid forest and forest edges, occasionally plantations; 2,950–5,900 ft (900–1,800 m)

SIZE Length: 2¾–3 in (7–7.5 cm). Weight: unknown

STATUS Critically Endangered

Peacock Coquette

This coquette is larger than other members of the genus and is uncrested, with shining, dark green plumage. The male has a white-spotted orange cheek patch and a generous ruff of broad, black-spotted, bright green "whisker" feathers, reminiscent of the "eye" markings on a peacock's train. The female has no ruff but has white-streaked cheeks. There are two subspecies: subsp. *pavoninus*; and subsp. *duidae*, which has almost no orange on the cheek. Feeding at all levels in the forest, this coquette is sometimes seen in groups but is regularly chased away by hummingbird species that are more territorial. It has a typical coquette flight action—rather slow and weaving, with the tail held out and tilted up. It floats smoothly up and down when searching foliage for insects. The cup-shaped nest straddles a branch. Incubation and fledging periods are 13 to 14 days and 20 days, respectively.

DISTRIBUTION Subsp. *pavoninus* occurs on Cerro Ptari-tepui and Mt. Roraima in southeastern Venezuela and the Merume Mountains in Guyana; subsp. *duidae* occurs on Mt. Duida and adjacent smaller peaks in southeastern Venezuela

HABITAT Rainforest, cloud forest, forest edges; 1,650–6,550 ft (500–2,000 m)

SIZE Length: 3¾ in (9.5 cm). Weight: unknown

STATUS Least Concern

Peruvian Piedtail

This is a rare hummingbird from the eastern slopes of the Andes in Peru. It has the same basic green-and-white patterning as the Ecuadorian Piedtail, but the lower belly is light orange rather than white, and there is also a little light orange on the throat sides. It feeds mainly at low levels, visiting shrubs, small trees, and epiphytes, and tends to cling rather than hover when taking nectar or picking small insects from the surrounding foliage. Very little is known about this hummingbird, but it has been classified as Near Threatened because its small range is undergoing rapid deforestation and the frequency of sightings suggest it is declining. It does, however, appear to be quite tolerant of habitats that have been modified to some extent, and is reported as being fairly common in some areas. Increasing the area of land under protection within its range would help curtail its decline.

DISTRIBUTION A small area of central and southeast Peru

HABITAT Submontane forest edges and secondary forest; 2,950–3,950 ft (900–1,200 m)

SIZE Length: 2¾–3 in (7–7.5 cm). Weight: 2–2.5 g

STATUS Near Threatened

Venezuelan Sylph

This hummingbird is confined to northeast Venezuela. The male has green and greeny-bronze underparts, with a shining green head and a bright blue throat patch. The long blue tail feathers are deep violet at their base, turning paler blue at the ends. The very different female is about half the length of the male; she is also green but has a blue head, and a mostly white throat and underparts. The female's tail is blue-green with white tips to the outer feathers, is forked, and is much shorter than the male's. The species feeds on invertebrates and nectar. Little studied, it is thought to build its nest from moss, with a side entrance and secured to a twig. It has been considered a subspecies of the Long-tailed Sylph and Violet-tailed Sylph, but is now generally recognized as a separate species. The Venezuelan Sylph is considered Endangered as its habitat is restricted and threatened by deforestation and development.

DISTRIBUTION Mountains of the Turimiquire Massif, northeast Venezuela

HABITAT Coastal slopes, subtropical and tropical forest slopes and edges, open scrubland; 4,750–6,000 ft (1,450–1,850 m)

SIZE Length: 3½–8¾ in (9–22 cm). Weight: 4.5–5.5 g

STATUS Endangered

Wedge-tailed Hillstar

A rather small, long-billed, colorful hillstar, the male of this species has the typical shining green throat of the genus, and a rich rufous belly with a central black stripe. The female's throat is white with dark spots, and her underside is lighter orange with no black stripe. The species visits a wide range of flowers for nectar and also hawks for small flying insects. Like other hillstars it shows a range of adaptations to life at high altitudes, including torpor when roosting. It constructs a substantial and well-lined cup nest which is fixed to a vertical surface. Its categorization as Near Threatened stems from its small population (less than 10,000 individuals) and ongoing decline due to habitat loss and degradation—it occurs in heavily populated areas where livestock grazing, planting of non-native species, and deliberate burning damage the native vegetation.

DISTRIBUTION A few mountainsides in southern Bolivia

HABITAT Scrubland and light woodland; 8,550–13,100 ft (2,600–4,000 m)

SIZE Length: 4⅜–5⅛ in (11–13 cm). Weight: 7.5–8.5 g

STATUS Near Threatened

Olivaceous Thornbill

A rather large and long-tailed thornbill, this species is almost entirely dark olive green, with a narrow iridescent throat patch that is green at the top and orange below. The female's throat patch is smaller and more diffuse but the sexes are otherwise alike. There are two subspecies: subsp. *olivaceum*; and the slightly smaller and lighter-colored subsp. *pallens*. This bird feeds at low levels and has even been observed on the ground visiting prostrate flowers. Insects make up a high proportion of its diet and it may be observed searching for them at ground level as well as hawking for flies. The male is territorial, defending concentrations of good nectar flowers such as *Castilleja*. The species seems to prefer to nest in *Polylepis* woodland. Though rather uncommon and localized, it has a stable population thanks to its remote habitat being relatively secure from development.

DISTRIBUTION Subsp. *olivaceum* occurs patchily in the Andes of southeast Peru and west-central Bolivia; subsp. *pallens* occurs patchily in the Andes of central Peru

HABITAT Puna grassland, scrub, woodland edges; 11,800–15,100 ft (3,600–4,600 m)

SIZE Length: 5½–5⅞ in (14–15 cm). Weight: 8–9 g

STATUS Least Concern

Perija Metaltail

A beautiful small hummingbird with a broad tail, short bill, and white "leg puffs," this species has a gleaming copper-violet tail and, in the male, a glittering green forehead and diamond-shaped throat patch. His plumage is otherwise blackish, while the female is a warm brown, paler below, with dark spotting on the belly. The species' behavior and breeding biology is presumed to be similar to those of the other metaltails, but its distribution means that it is very little studied—the areas where it occurs are known for illegal activities, including the cultivation and smuggling of narcotics. Its status of Endangered (upgraded from Vulnerable in 2004) reflects its ongoing decline and the precarious status of its habitat. One of its sites, Cerro Tetari, is nominally protected but at present is not being actively managed for conservation of this or other threatened species.

DISTRIBUTION Serranía del Perijá, northern Colombia and Venezuela

HABITAT Scrub, elfin forest, forest edges; 9,200–10,500 ft (2,800–3,200 m)

SIZE Length: 3⅞–4⅜ in (10–11 cm). Weight: 3.5–4 g

STATUS Endangered

Fire-throated Metaltail

The male Fire-throated Metaltail has a red throat patch, outlined in shining green. The female has a much smaller red patch with no iridescent border, and her belly is whitish with green streaks. Both sexes are otherwise bright olive green. The species is short-billed, and like its relatives it has small puffs of white feathers at the base of its legs. This bird is very closely related to the other small green *Metallura* metaltails and shares their habits, including the male's territoriality, and a preference for feeding from flowers of the family Melastomataceae. It is less specialized than some of the other metaltails though, and therefore its prospects for the future are of less cause for concern. However, it does have a quite restricted distribution and its habitat is under some pressure from human activities such as burning to clear land for cattle ranching.

DISTRIBUTION A small area on the eastern slopes of the Andes in central Peru

HABITAT Shrubby glades in damp, mossy, stunted forest and adjoining paramo; 9,500–13,100 ft (2,900–4,000 m)

SIZE Length: 4⅜ in (11 cm). Weight: 5 g

STATUS Least Concern

Buff-thighed Puffleg

A blue-tailed green puffleg, this species has substantial buff-tinged white leg puffs at the leg bases, and its body plumage is greener than that of the otherwise very similar but geographically separated Greenish Puffleg, with just a hint of yellowish on the rump. The sexes are similar. Two subspecies are recognized: subsp. *assimilis*; and subsp. *affinis*, which has warm brown leg puffs. This is a territorial feeder, defending clumps of low-growing flowering bushes from intruders. It feeds while hovering, with the tail slightly cocked. It also picks insects and spiders from foliage but seldom flycatches in flight. Little is known of its breeding behavior but presumably it builds a ball-shaped moss nest like others of its genus. There is evidence that it undertakes seasonal altitudinal movements. The nominate subspecies is quite common, but subsp. *affinis* has been recorded at only a handful of sites.

DISTRIBUTION Subsp. *assimilis* occurs from southern Peru to northwest Bolivia; subsp. *affinis* occurs in the eastern Andes of north-central Peru

HABITAT Dense, wet forest and forest edges on lower slopes; 4,900–9,850 ft (1,500–3,000 m)

SIZE Length: 3½–3⅞ in (9–10 cm). Weight: 5–6 g

STATUS Least Concern

Gorgeted Puffleg

The male of this puffleg has a glittering violet-blue throat and vent, and shining green throat sides. It is otherwise blackish with slight green iridescence. The female is lighter green, and her throat patch is smaller. Both sexes have sizable white leg puffs. This species was discovered only in 2005, and officially described to science in 2007. Just nine birds (seven of them males) have been mist-netted by researchers, and the area of habitat within which they were caught is less than four square miles (10 km²). It can therefore be inferred that the population is tiny, and is likely to decline as habitat destruction is ongoing as the dwarf forest is cleared to make way for farmland and illegal coca plantations. Protecting the species' remaining habitat is obviously essential to save it but local issues make this a difficult objective. However, a multipronged conservation effort is underway.

DISTRIBUTION Serranía del Pinche, southwest Colombia

HABITAT Stunted cloud forest; 8,550–9,500 ft (2,600–2,900 m)

SIZE Length: 3¾–3⅞ in (9.5–10 cm). Weight: 4–4.5 g

STATUS Critically Endangered

Turquoise-throated Puffleg

A medium-sized, glittering green puffleg, this species has a blue chin patch (absent in the female) and violet-blue vent. The female's underparts have some white speckling and a yellow tint. This is one of the world's "missing" species of hummingbirds, as it has not been recorded with certainty since the nineteenth century. Of the six museum specimens, location details are known for only one, and the natural vegetation at the site where it was taken is now entirely destroyed. There is therefore a strong possibility that the species has since become extinct. However, a fairly recent unconfirmed sighting in 1976 and the existence of remnant patches of similar habitat mean that it may still hang on, and thorough surveying is needed. Nothing is known of its habits, but it is highly likely to be similar to those of the other species in its genus in most respects.

DISTRIBUTION Northwest Ecuador

HABITAT Forest and scrub at forest edges; 6,900–7,550 ft (2,100–2,300 m)

SIZE Length: 3⅞–4⅜ in (10–11 cm). Weight: unknown

STATUS Critically Endangered

Blue-capped Puffleg

Named for the male's deep blue forehead patch, bordered with lighter blue, this is a small and particularly lovely puffleg. The female lacks the cap, and has an unmarked light orange-buff throat and breast, while the male's are green with a touch of blue. Both sexes are otherwise glistening, deep green with dark wings and deep blue vents and tails. It is a rather uncommon species, though not thought to be declining, and occurs farther south than other members of the genus. It takes nectar from flowers low in the vegetation and is very active and quick when feeding, driving other hummingbirds from its chosen patch. Its breeding behavior is largely unknown; it probably nests in well-sheltered spots, as its habitat is subject to heavy rainfall. It makes altitudinal movements according to the seasons (wet or dry) in the south of its range at least.

DISTRIBUTION Central and southeast Bolivia into northwest Argentina

HABITAT Humid scrubby areas, forest edges; 4,900–9,500 ft (1,500–2,900 m)

SIZE Length: 3½–3⅞ in (9–10 cm). Weight: 4–4.5 g

STATUS Least Concern

Colorful Puffleg

The male of this small puffleg is an exquisite multicolored bird—iridescent green on the head and upperside, shading to deep blue on the belly, with a scarlet vent and yellow undertail. The substantial leg puffs are white. The plainer female is green with a white belly, heavily marked with green spotting on the throat and breast, which changes to reddish spotting on the belly. It has a relatively short, straight bill. This species feeds at medium heights in the forest strata, taking nectar from flowers such as *Miconia* and *Clusia*. It is little-known, being discovered only in 1967 and then not observed reliably again until 1997. It occurs at four known locations but is unpredictable and difficult to find at all of them, and its estimated population is no more than 1,500 individuals. Habitat loss in its known and potential sites is ongoing, but a reserve has recently been established.

DISTRIBUTION Western Andes in Colombia

HABITAT Forest interior, occasionally forest edges; 7,200–7,850 ft (2,200–2,400 m)

SIZE Length: 3⅛–3½ in (8–9 cm). Weight: unknown

STATUS Critically Endangered

Violet-throated Starfrontlet

This variable hummingbird is primarily olive green, shading to orange-gold on the belly, rump, and tail. The male has a dark head with a shining violet throat patch and a forehead patch that varies according to the subspecies. The four known subspecies are subsp. *violifer*, which is darkest and has a thin whitish breast band; subsp. *dichroura*, which is greener, with more dark in the tail, and has a bright green forehead patch; subsp. *albicaudata*, which has a blue forehead patch and white tail sides; and subsp. *osculans*, which has a blue forehead patch. It is a low-level trapliner, sometimes moving along paths or trails within the forest. It moves to lower altitudes as it follows the flowering season of montane plant species, and males and females tend to occur at different levels when they are not breeding. Because it does not readily use modified habitats, the species is at risk from deforestation.

DISTRIBUTION Subsp. *violifer* occurs in the eastern Andes of northwest Bolivia; subsp. *dichroura* occurs in south Ecuador and the eastern Andes of north and central Peru; subsp. *albicaudata* occurs in the Apurimac region of southern Peru; subsp. *osculans* occurs in the eastern Andes of southeast Peru

HABITAT Open areas and cloud and elfin forest edges; 4,250–12,150 ft (1,300–3,700 m)

SIZE Length: 5⅛–5¾ in (13–14.5 cm). Weight: 5.5–11.5 g

STATUS Least Concern

Pink-throated Brilliant

This small brilliant hummingbird has a dark tail and wings and a white undertail. The male has shining, dark pink throat sides, while the female has just a touch of pink and some dark spotting on the breast. The species is very little known despite its sizable range. It has been observed feeding from *Psittacanthus* and related mistletoes that grow high in the forest strata, and its feeding and breeding behaviors are presumably similar to those of the other *Heliodoxa* species. Previously considered to be Near Threatened, its status was changed recently in 2012 to Vulnerable because of the likely rate of continued deforestation in its habitat. Its population is not known but the bird is usually reported to be rare or very rare. Urgent assessment of the species' population size and exact distribution will help to guide future conservation efforts, but its forest habitats are under tremendous pressure.

DISTRIBUTION Foothills of the east Andes in south Colombia, northeast Ecuador, north Peru

HABITAT Humid montane forest; 1,300–2,950 ft (400–900 m)

SIZE Length: 4⅜–4¾ in (11–12 cm). Weight: unknown

STATUS Vulnerable

Rufous-webbed Brilliant

This species is very similar to the Pink-throated Brilliant, especially the male (the female has more heavily spotted underparts than the female Pink-throated), but can be distinguished by its distribution, which does not overlap with that of the Pink-throated. The plumage is primarily dark, shining green, with the male's forehead being particularly strongly iridescent. The bird is named for the reddish bases to the inner-tail feathers, a feature most evident when it is hovering. Studies of this species' ecology are so far almost nonexistent, although it can be inferred that it has much in common with the other members of the genus. It is known to be declining, and because its distribution is patchy there are likely to be some very small, isolated populations that are particularly vulnerable. However, the fact that the brilliant is sometimes seen in plantations suggests that its habitat needs are not too exacting.

DISTRIBUTION Central and southern Peru

HABITAT Forest edges, woodland, sometimes forest interiors, plantations; 2,450–4,250 ft (750–1,300 m)

SIZE Length: 4⅜–4¾ in (11–12 cm). Weight: 5–6 g

STATUS Least Concern

Golden-crowned Emerald

This species was originally classed as a subspecies of the Blue-tailed Emerald, then Canivet's Emerald, but is now recognized as a species in its own right. It is a long-tailed emerald, similar to the Cozumel Emerald (with which it is sometimes grouped), but the tail feathers are narrower and the bill shorter. The green plumage has a yellowish-golden sheen. As with others in the genus, the female's underside is whitish-gray. Like other emeralds, this is a lone feeder and a trapliner that often avoids the attentions of more dominant, territorial species by choosing flowers too small to be of interest to them. The nest is a small cup built very low in a shrub. The chicks fledge at 24–25 days, a longer fledging period than that reported for the Blue-tailed Emerald. The population dynamics for this species are not well understood.

DISTRIBUTION Western Mexico

HABITAT Dry tropical forest; 0–5,900 ft (0–1,800 m)

SIZE Length: 2½–3⅜ in (6.5–8.5 cm). Weight: 3–3.5 g

STATUS Least Concern

Cozumel Emerald

This is a vivid, glossy green hummingbird with dark wings and a relatively long, deeply forked dark tail, which is much longer in males than females. Females also have light gray rather than shining green underparts, and dark ear coverts with a small post-ocular white spot. This species was formerly considered to be a subspecies of Canivet's Emerald (which was itself split from the Blue-tailed Emerald), but it was designated a full species in the late 1990s. The Cozumel Emerald feeds mainly rather low down, visiting smaller flowers but also possibly piercing the corollas of larger ones to steal nectar. In addition, it hawks for insects. Though it has a very limited distribution of just one island off the coast of Mexico, this hummingbird is quite numerous and is not thought to be a species of conservation concern.

DISTRIBUTION Cozumel island, off the coast of Yucatán, Mexico; also reported from nearby Holbox and Mujeres islands, but it does not appear to have established populations there

HABITAT Scrub, forest; 0–45 ft (0–14 m)

SIZE Length: 2½–3⅜ in (6.5–8.5 cm). Weight: 3–3.5 g

STATUS Least Concern

Chiribiquete Emerald

This fairly large emerald has a long, straight bill. The male is shining green with a diffuse blue throat patch and a dark tail and wings, while the female has a white belly and a greener tail. It is similar to the Blue-tailed Emerald and may simply be a subspecies of this bird. The species' favorite nectar plant is *Decagonocarpus cornutus*, but it also visits other shrubs and small trees, apparently feeding by traplining rather than holding a territory. It is an enthusiastic flycatcher and also gleans small insects from foliage, especially from *Bonnetia* shrubs. This species is well adapted to the unique rugged and diverse landscape of the Serranía de Chiribiquete in southeast Colombia, and is the most common hummingbird in the Chiribiquete National Park (the country's largest). Its habitat is not threatened as most of it is protected, and in any case the rocky terrain is unsuited to agriculture, so although the species' population trend is unknown it is likely to be stable.

DISTRIBUTION Serranía de Chiribiquete, southeast Colombia

HABITAT Open scrub, savanna, and into forest edges; 0–3,300 ft (0–1,000 m)

SIZE Length: 3⅜–3½ in (8.5–9 cm). Weight: 3.5–4 g

STATUS Least Concern

Brace's Emerald

This island jewel has the sad distinction of being one of only two hummingbirds confirmed to have become extinct in historic times. It must have been already very rare in 1877, when Lewis J. K. Brace collected the only known specimen, a male, on New Providence Island near the town of Nassau. For over a century it was considered a subspecies of the Cuban Emerald, until the discovery of Pleistocene-era hummingbird fossils on New Providence encouraged ornithologists to reexamine the specimen. Brace's Emerald was smaller than the Cuban Emerald, with feathers extending farther down its bill. The male's plumage was predominantly bronzy green, with a bright green crown and a golden sheen on the back. The female probably resembled females of the Cuban Emerald. The cause of the species' extinction is unclear, but it may have involved natural climate change, habitat disturbance caused by human settlers, or some combination of factors.

DISTRIBUTION Formerly Nassau, Bahamas

HABITAT Undisturbed dry evergreen forest (coppice)

SIZE Length: male 3¾ in (9.5 cm). Weight: unknown

STATUS Extinct

Caribbean Emerald

Virtually nothing is known about the extinct Caribbean or Gould's Emerald, not even its former home. The species was described by John Gould from the skin of a single adult male. No locality data accompanied the specimen, but it is thought to have come from either Jamaica, which has no other members of this genus, or the northern Bahamas, which are currently home to the Cuban Emerald. The male Caribbean Emerald was similar to the male Cuban Emerald, though with darker plumage and a shorter, more shallowly forked tail. The underparts and head were deep golden green, shading to bronze on the back, coppery purple on the uppertail coverts, and dark purple on the tail. The female was probably similar to other female emeralds, with pale underparts and outer tail feathers banded in green, black, and white. As with many other island species, habitat disturbance by human settlers and introduced predators are likely causes of its extinction.

DISTRIBUTION Unknown, possibly Jamaica or northern Bahamas

HABITAT Unknown

SIZE Length: male 3⅞ in (10 cm). Weight: unknown

STATUS Extinct

Hispaniolan Emerald

The male Hispaniolan Emerald is dark green above, with dark brown on the head and forehead, a shining green throat with a black patch on the breast, and dark green and brown underparts. The brown tail has a marked fork. The short bill is downcurved and has a pink lower mandible, turning black towards its tip. The female is similar to the male, but has a brown head, is darker green above, has a gray throat and breast, turning darker underneath, has white tips to the tail, and has a more downcurved bill. The species is a trapliner, usually feeding at low levels on the nectar of a wide range of plants, including *Heliconia bihai*, *Rhytidophyllum auriculatum*, *Aechmea*, and *Hibiscus*; it also catches small insects. The bird builds a small cup-like nest from plant fibers, moss, and lichen. The current global population and population trend are unknown but appear stable, and the species is regarded as common throughout its range.

DISTRIBUTION Hispaniola

HABITAT Dense forest, forest edges, scrubland; 650–8,200 ft (200–2,500 m)

SIZE Length: 3⅛–4⅜ in (8–11 cm). Weight: 2–5 g

STATUS Least Concern

Puerto Rican Emerald

The male Puerto Rican Emerald is dark, shining green above, with bright green on the head and forehead, a shining blue-green throat, and green underparts. The bright, shining blue tail has a pronounced fork. The lower mandible of the bird's short, straight bill is red, turning black towards its tip. The female is very similar to the male, but the bird is a paler green above, the head and forehead are dark green, the throat and breast are pale gray, turning darker underneath, the tail has a less pronounced fork, and the bill is all black. A trapliner, the species usually feeds at low levels on the nectar of a wide range of plants, including *Hohenbergia portoricensis*, *Vriesea sintenisii*, and *Epidendrum*; it also catches small insects and spiders. The bird builds a small cup-like nest from plant material, including plant fibers, lichen, and wild cotton. The current global population and population trend are unknown but appear stable, and the species is regarded as common throughout its range.

DISTRIBUTION Puerto Rico

HABITAT Mountainous areas, open forest, plantations, coastal mangroves; 0–3,300 ft (0–1,000 m)

SIZE Length: 2¾–3⅞ in (7–10 cm). Weight: 3–4 g

STATUS Least Concern

Coppery Emerald

Though unmistakably an emerald, the small Coppery Emerald lives up to its name and is distinct from its congeners thanks to the strong coppery suffusion—strongest on the back and tail—to its green plumage. The white-bellied female also has copper plumage tones, though they are less pronounced than in the male. This bird is a trapline feeder with the typical rapid, darting and direct flight of its genus, and usually visits flowers at heights above 13 feet (4 m). In contrast, the nest is built rather low down (usually at about 5 feet/1.5 m above ground) on top of a sloping branch. The female incubates the two eggs for 15 to 16 days and the chicks fledge at about 20 days. This is a rather scarce bird although its range is quite large. Its versatility and willingness to use degraded and modified habitats make it less vulnerable to decline. It exhibits seasonal altitudinal movements, usually following the reproductive season.

DISTRIBUTION Northeast Colombia, northwest Venezuela

HABITAT Scrubland, woodland edges, plantations, parkland, farmland; 0–8,550 ft (0–2,600 m)

SIZE Length: 2¾–3⅜ in (7–8.5 cm). Weight: 3–3.5 g

STATUS Least Concern

Narrow-tailed Emerald

The male of this typical emerald has shining green plumage with dark flight feathers, while the female has white underparts with a strong greenish-brown suffusion on the flanks. There are two subspecies, subsp. *stenurus* and subsp. *ignotus*, the latter being smaller and with yellower plumage and a darker tail. This bird is a trapliner that prefers to feed in more open areas, visiting a wide range of flower types, typically up to 13 feet (4 m) above ground. It also hawks insects in flight. The nest is built low in a shrub or small tree, and is made mainly of moss with an outer layer of lichen. The incubation period is 15 to 16 days, with chicks fledging after a further 20 days. Birds may move to higher altitudes after breeding. The population trend is not known but this is a common bird in much of its range.

DISTRIBUTION Subsp. *stenurus* occurs in northwest Venezuela and northeast Colombia; subsp. *ignotus* occurs in northwest Venezuela from the coastal range to the extreme southwest of Lara state

HABITAT Damp forest and secondary growth, scrubland; 3,300–9,850 ft (1,000–3,000 m)

SIZE Length: 3–3½ in (7.5–9 cm). Weight: 3–3.5 g

STATUS Least Concern

Dusky Hummingbird

This is a dark hummingbird with subdued plumage tones. It is olive green above and gray-green below, with dark ear coverts and a narrow white stripe behind the eye. The sexes are similar but the male is a little paler and has a black-tipped red bill (the female's bill is darker). This hummingbird takes nectar mainly from flowering trees and catches insects around flowering cacti, preferring to feed at high levels. It tends to fan out and bob its tail repeatedly while it hovers. The nest is built low in a shrub and is a small cup, lined inside with fluff from plant seeds and a few feathers. At a glance, it resembles a collection of natural debris as it is heavily camouflaged with plant matter such as dead grass and bark fragments. Dusky Hummingbirds are common across their range and frequently visit gardens.

DISTRIBUTION South Mexico

HABITAT Dry open countryside with some trees and shrubs, gardens; 2,950–7,200 ft (900–2,200 m)

SIZE Length: 3½–3⅞ in (9–10 cm). Weight: 4.5 g

STATUS Least Concern

Blue-headed Hummingbird

Emerald-chinned Hummingbird

This is a colorful, richly toned, medium-sized hummingbird. The male is dark, shining green, shading to blue on the head, while the female is green with white underparts and dark ear coverts. Though it can be seen at sea-level, this bird is more common on higher ground above 2,600 feet (800 m). It visits flowers from ground level to the canopy, and flycatches over forest streams as well as gleaning arthropods from foliage. The female builds a nest 3 to 13 feet (1 to 4 m) above ground on a level twig or fern frond, using fluff from silk cotton tree (*Bombax*) seeds as a lining. The eggs hatch at 16 to 18 days and the chicks fledge at 20 to 23 days, though they remain with the female for up to 28 days after that. The population suffered severe losses on Dominica following hurricanes in the 1980s but is now recovering.

DISTRIBUTION Dominica, Martinique

HABITAT Forest, forest edges, and secondary growth along rivers; 0–3,300 ft (0–1,000 m)

SIZE Length: 3½–4⅜ in (9–11 cm). Weight: 4.5 g

STATUS Least Concern

This is a small, neatly marked hummingbird with a short bill. The male is dark, shining green with a vivid, glistening green throat patch that is offset by a dark breast and ear coverts, while the female's underside is white. There are two subspecies, subsp. *abeillei* and the smaller, more golden-toned subsp. *aurea*. The Emerald-chinned Hummingbird forages in the forest understory, visiting flowers of the families Rubiaceae, Verbenaceae, and Onagraceae, which the male defends from other hummingbirds. The nest is a rather deep, thick-walled cup made of soft plant materials, but little more is known of the bird's breeding behavior. Like other species dependent on primary forest, it has been impacted by logging, especially in Mexico. The population is declining, with an estimated maximum of just 50,000 individuals..

DISTRIBUTION Subsp. *abeillei* occurs in southeast Mexico through Guatemala and El Salvador to north Honduras; subsp. *aurea* occurs in south Honduras and north Nicaragua

HABITAT Cloud forest, occasionally forest edges; 3,300–7,200 ft (1,000–2,200m)

SIZE Length: 2¾–3 (7–7.5 cm). Weight: 2.5 g

STATUS Least Concern

Long-tailed Sabrewing

This sabrewing, formerly considered a subspecies of the Wedge-tailed Sabrewing, has a sturdy bill and long, wedge-shaped tail. The plumage is shining green above with a light blue crown, and whitish gray below. The sexes are similar, though males are larger on average than females. The species has been observed feeding from *Heliconia, Tillandsia dasyliriifolia*, and Tiger Tree (*Erythrina folkersii*) flowers, among others, and is territorial around flower clusters. Males form singing leks to attract mates, gathering in twos or threes to sing and display from perches. During the display, the males fan and waggle their tails. This species has a very restricted distribution, and shows little tolerance of modified habitats. Because its forest habitat is being cleared for agriculture and ranching, the species is likely to be declining. Surveys are needed to build a clear picture of its population and to study its ecology so that suitable habitat can be identified and protected.

DISTRIBUTION South Mexico, farther north than the Wedge-tailed Sabrewing

HABITAT Forests, forest edges and clearings, secondary growth; 0–4,250 ft (0–1,300 m)

SIZE Length: 4¾–5½ in (12–14 cm). Weight: 6–8 g

STATUS Near Threatened

Rufous-breasted Sabrewing

Also known as the Tepui Sabrewing, this species is found only on isolated table mountains (tepui) of Venezuela and Brazil; some taxonomists consider it to be the same species as the Buff-breasted Sabrewing. The sexes are similar, with shining green upperparts, a green head, a pale orange face and cheeks, and pale orange-red underparts. The medium-length black bill is very slightly downcurved, and there are white spots behind each eye. The central tail feathers are bronze and the outer tail feathers pale orange-red. Rufous-breasted Sabrewings typically feed at low levels in the understory on the nectar of many different plants, as well as small insects. Details of the species' breeding behavior are unknown. The current global population is also unknown, but although the Rufous-breasted is found only in restricted areas of Venezuela and Brazil, these populations are thought to be relatively stable, and the species is recorded as fairly common locally.

DISTRIBUTION Southeastern Venezuela, northwestern Brazil

HABITAT Mountain areas, low tepui, forests, scrublands; 3,300–8,200 ft (1,000–2,500 m)

SIZE Length: 3⅞–5⅛ in (10–13 cm). Weight: 5–7 g

STATUS Least Concern

White-tailed Sabrewing

The male of this large hummingbird is shining green above and below, with a shining, dark blue-purple throat patch. It has a medium-length, downcurved black bill, and small white spots behind each eye. The central feathers of the square tail are dark green-black, and the outer tail feathers have large white tips. The female is similar but is gray-green below. The White-tailed Sabrewing typically feeds at low and mid-levels on the nectar of many different plants, including *Heliconia*, Rubiaceae, bromeliads, and *Musa* (banana), and catches small insects by hawking. It uses plant fibers and moss to build a sturdy cup-like nest, which it builds in branches overhanging water or the ground. The current global population has not been measured, but is thought to be in decline. Although the species is considered to be fairly common in its range, its restricted habitat is threatened.

DISTRIBUTION Northeastern Venezuela, Tobago

HABITAT Mountainous areas, forests, forest edges, plantations, stream banks, gardens; 1,000–6,550 ft (300–2,000 m)

SIZE Length: 4⅜–5½ in (11–14 cm). Weight: 6–9 g

STATUS Near Threatened

Santa Marta Sabrewing

This is a typical sabrewing, the male showing thickened, curved shafts to the outer primary feathers that give the folded wing a distinctive shape. The bill is rather sturdy and downcurved. The male's plumage is dark, glossy green with a shining, dark blue throat and upper breast; the female has whitish underparts. This species feeds at low levels from banana (*Musa*) flowers (no other nectar sources are documented), and on insects. It descends to lower elevations in the dry season, and in the wet season moves up almost to the snowline to breed. It has a tiny range, and an estimated population of between 1,500 and 7,000 individuals. Habitat loss is ongoing at lower levels in the Sierra Nevada de Santa Marta, as forest is cleared to allow cultivation, including of illegal drugs, despite nominal protection (the area is a UNESCO Biosphere Reserve).

DISTRIBUTION Sierra Nevada de Santa Marta, northeast Colombia

HABITAT Humid forest edges, old plantations; 3,950–15,750 ft (1,200–4,800 m)

SIZE Length: 5⅛ in (13 cm). Weight: unknown

STATUS Endangered

Buff-breasted Sabrewing

Very closely related to the Rufous-breasted Sabrewing and sometimes thought to be the same species, the Buff-breasted Sabrewing is found only on isolated table mountains (tepui) of Venezuela and Brazil. The sexes are similar to one another and to the Rufous-breasted, with shining green upperparts, a green head, and a pale orange face and cheeks, but with much duller underparts than the Rufous-breasted. The medium-length black bill is very slightly downcurved, and birds have white spots behind each eye. The central feathers are bronze, and the outer tail feathers pale orange-red. Subsp. *guaiquinimae* has darker underparts and tail than the nominate. The species typically feeds in the understory and at mid-levels in the forest on the nectar of many different plants; it also takes small insects. Details of its breeding behavior are unknown. The current global population is also unknown, but it is thought to be relatively stable.

DISTRIBUTION Subsp. *duidae* occurs in Venezuela and northern Brazil; subsp. *guaiquinimae* occurs in south-central Venezuela

HABITAT Mountain areas, low tepui, forest, scrublands; 3,300–8,200 ft (1,000–2,500 m)

SIZE Length: 3⅞–5⅛ in (10–13 cm). Weight: 5–7 g

STATUS Least Concern

Blue-capped Hummingbird

Also known as the Oaxaca Hummingbird because of where it is from, this endangered hummingbird is a striking medium-sized species with a rather short, square tail and long, fine bill. The male is vivid green with broad white tail sides, a chestnut wing patch, and a dark blue crown. The female is drabber, with mottled green upperparts and a gray-white underside, and the wing patch is smaller and browner. This is not a particularly territorial species, and readily feeds at all heights in the forest strata. The few observed nests were cups placed on branch forks, usually rather low down in a small tree. Known from only two sites in the Sierra Miahuatlán mountains, the Blue-capped Hummingbird has a population of some 1,700 individuals and is continuing to decline. Deforestation of its habitat, which began in the 1960s, is the main cause. However, much habitat was also destroyed by Hurricane Pauline in 1997.

DISTRIBUTION South-central Mexico (Oaxaca)

HABITAT Cloud forest, gallery forest, small forest clearings and edges; 2,300–8,550 ft (700–2,600 m)

SIZE Length: 3⅞–4⅜ in (10–11 cm). Weight: 4.5–5 g

STATUS Endangered

White-tailed Hummingbird

A fairly typical *Eupherusa* species, the White-tailed Hummingbird has more extensive white in its tail sides than its congeners. Both sexes have a prominent chestnut wing patch; the male is otherwise green with dark wings; the female has a gray-white underside. This species visits a wide range of flower types, including *Inga*, *Lobelia*, and *Psittacanthus*, but is not territorial and is regularly chased off by other similar-sized hummingbirds such as the *Amazilia* species. Its status was revised from Endangered to Vulnerable in 2000, as researchers established that it was present at five locations and was quite common there. However, it is still declining, and now has an estimated population of about 15,000 individuals. Its habitat is highly pressured—one of its known sites is protected, but further protection of at least one of the other sites is urgently needed.

DISTRIBUTION South Mexico (west Oaxaca)

HABITAT Cloud forest and other wooded habitats; 2,950–7,850 ft (900–2,400m)

SIZE Length: 3⅞–4⅜ in (10–11 cm). Weight: 4.5–5 g

STATUS Vulnerable

Long-tailed Woodnymph

The male Long-tailed Woodnymph has a shining green head and neck, a brilliant green throat, shining purple-blue underparts, and a purple-blue back. The medium-sized black bill is slightly downcurved and the long blue tail has a marked fork. The female has a duller green head, a blue-green back, and gray underparts. Her green tail is shorter than the male's and less forked, with a broad blue band and white tips on the outer tail feathers. The Long-tailed feeds on the nectar of a wide range of flowers, including Bromeliaceae, Passifloraceae, Rubiaceae, and Verbenaceae, as well as on small insects. It builds a cup-like nest of plant fibers and moss, usually on a branch protected by overhanging leaves. The current global population is unknown, and the species is regarded as common, although it is thought to be declining as a result of habitat loss.

DISTRIBUTION Eastern Brazil

HABITAT Rainforest near coasts, forest edges, cerrado, semi-open areas, plantations, gardens; 0–1,950 ft (0–600 m)

SIZE Length: 3½–5⅛ in (9–13 cm). Weight: 4–7 g

STATUS Near Threatened

Honduran Emerald

This is a rather plain *Amazilia* hummingbird, with drab olive-green plumage above and a dusky gray breast and belly. The male has a broad, glittering turquoise throat patch, while the female's throat is whitish with heavy green spotting. This species feeds at both low and high levels in the strata, visiting flowers of *Pithecellobium*, *Pedilanthus*, and cacti. It is Honduras's only endemic bird. Between 1950 and 1988, the species was not observed at all, but was then discovered at two sites. A third site was discovered in 2007, leading to the bird's conservation status being downgraded from Critically Endangered to Endangered. However, the species remains severely threatened and is declining, with a population estimated at 350 to 1,500 individuals. Clearance of land for ranching and other development is the main threat it faces. Part of its habitat is now protected and there are plans to expand the area under protection.

DISTRIBUTION North and central Honduras

HABITAT Dry thorny forest, scrubland, light woodland; 250–1,000 ft (75–300 m)

SIZE Length: 3½–3⅞ in (9–10 cm). Weight: unknown

STATUS Endangered

Green-fronted Hummingbird

Formerly classed with the *Argyrtia* emeralds, this is a rather dull green hummingbird, brightest on the forehead, with white underparts, a copper-tinged tail, and a black-tipped red bill. Subsp. *wagneri* differs from the nominate by having a bronze tinge around the lower back, rump, neck sides, and breast sides; while subsp. *villadai* has less green on the flanks and sides. The species may be seen feeding at all heights in the forest strata but most commonly at mid- to high levels, where it takes nectar and insects. The nest is built quite high in a tree or bush and is made mainly from soft plant fibers and cobwebs, with lichens added to the outer layer for camouflage. After breeding, some birds move to lower elevations. Although the species' conservation status is Least Concern, under Mexican law it is considered threatened because it has a small range and its main habitat is at risk.

DISTRIBUTION Subsp. *viridifrons* occurs in south Mexico (western Oaxaca to central Guerrero); subsp. *wagneri* is found in central and south Oaxaca only (south Mexico); subsp. *villadai* occurs in southeast Mexico (southeast Oaxaca and Chiapas)

HABITAT Dry, open pine and pine–oak forest, thorn forest, gardens, parkland; 150–4,600 ft (50–1,400 m)

SIZE Length: 3⅞–4½ in (10–11.5 cm). Weight: 5.5 g

STATUS Least Concern

Black-billed Streamertail

This species replaces the Red-billed Streamertail in the east of Jamaica, and closely resembles it. The most obvious difference is that the male's bill is black rather than red, and the female has a grayish-black bill rather than the dull pinkish-brown bill of the female Red-billed. In addition, the two species have significantly different vocalizations and display behavior; prior to studies that revealed these differences in 1999, the two were considered to belong to the same species. The Black-billed Streamertail is a common bird in most parts of its range, and takes nectar from many different native and introduced flower species, as well as insects caught in flight and picked from foliage. It nests at all times of year, the eggs hatching at 17 to 19 days and chicks fledging at 17 to 19 days (although they remain dependent on the female for another three to four weeks).

DISTRIBUTION Eastern tip of Jamaica

HABITAT Humid, well-vegetated habitats, including elfin forest, gardens, and parks; 0–3,300 ft (0–1,000 m)

SIZE Length: 4⅛–13¾ in (10.5–35 cm). Weight: 4.5 g

STATUS Least Concern

Pirre Hummingbird

The male of this fairly small hummingbird species is mainly shining green with blackish wings. There is a reddish patch in front of the eye, spreading to the front of the forehead and chin, and a chestnut patch on the secondaries. The outer tail feathers are chestnut. Females have light peachy-buff underparts. A little-known species of the forest interior, the Pirre Hummingbird visits flowers at low and medium heights in the forest strata, and can be seen around plants of the families Ericaceae and Rubiaceae. It is designated Near Threatened because it appears to have an extremely limited natural range. The forests it occupies are largely untouched, so although its population trend is not known there is reason to believe it is fairly secure for the time being. However, proposed developments in the area are likely to impact seriously on this species.

DISTRIBUTION Eastern Panama, far northwest Colombia

HABITAT Humid forest and forest edges; 1,950–4,900 ft (600–1,500 m)

SIZE Length: 3⅜–3¾ in (8.5–9.5 cm). Weight: 3–4 g

STATUS Near Threatened

Violet-capped Hummingbird

Rather similar to the Pirre Hummingbird in general appearance and closely related to that species, this hummingbird is bright, glistening green with (in the male) a deep violet cap that reaches just short of the nape. The wings and tail are dark. Females have green crowns and grayish underparts, and have some green spotting on the flanks and sides of the breast. Both the Violet-capped and Pirre Hummingbirds have unusually long, stiffened undertail feathers. This hummingbird usually forages alone, traplining a low-level route through the understory, where it takes nectar from shrubs such as *Salvia* and *Palicourea*. It will also visit hummingbird feeders. Its behavior is not yet well studied, but it is thought to be sedentary, and it is known to breed between December and March. The forest habitat it occupies is mainly pristine and it is a fairly common species in much of its restricted range.

DISTRIBUTION Central Panama, extending just into northwest Colombia

HABITAT Moist, dense forest and forest edges; 1,950–3,950 ft (600–1,200 m)

SIZE Length: 3⅜–3¾ in (8.5–9.5 cm). Weight: 3.5–4.5 g

STATUS Least Concern

Humboldt's Sapphire

This species was formerly classed as a subspecies of the Blue-headed Sapphire, but was split from it in 2005. The male is almost entirely bright, shining green, shading to blue on the chin, forehead, and tail. The wings are dark. The female has a white underside with some green spotting on the breast sides and flanks. This newly designated species is distinct from the Blue-headed Sapphire in ecology as well as appearance—the latter occurs at much higher elevations, while Humboldt's Sapphire is restricted to low-lying coastal areas, where it feeds preferentially from *Pelliciera rhizophorae* flowers and is quite territorial around good nectar sources. Humboldt's has a small natural range and its habitat is affected by the cutting of mangrove trees for firewood and to clear space for shrimp farming. Although currently classed as Least Concern, the species is rather rare and probably declining.

DISTRIBUTION Southeast Panama to northwest Ecuador

HABITAT Mangroves and wet forest on the coast; 0–150 ft (0–50 m)

SIZE Length: 4⅜ in (11 cm). Weight: 6.5–6.8 g

STATUS Least Concern

Violet-chested Hummingbird

A medium-sized hummingbird with a long, gently downcurved bill, this species is the only representative of its genus, but it is closely allied to the brilliants (genus *Heliodoxa*), which it resembles apart from its longer, more downcurved bill. It is mostly shining green with a rufous patch on the secondaries. The male has a strongly iridescent green throat and violet breast patch, while the female has pale buff underparts, heavily speckled with green. This is a species of the dark forest interior, feeding among low thickets of *Heliconia* and in other well-shaded spots. It is solitary and fiercely territorial, and spends its resting time perched close to its favored clump of flowers, ready to chase away any rivals. Quite common in most of its range, though scarce in the Andes, its ability to use secondary growth gives it some protection from loss of its primary habitat. It is present in some protected areas.

DISTRIBUTION Andes and coastal mountains of northern Venezuela

HABITAT Humid woodland, forests, and secondary growth; 0–6,250 ft (0–1,900 m)

SIZE Length: 4¾–5⅛ in (12–13 cm). Weight: 8.5–10.5 g

STATUS Least Concern

Plain-capped Starthroat

The Plain-capped Starthroat is a medium-sized hummingbird of the Pacific slope of Mexico and Central America. The most conspicuous markings on its understated plumage are broad white mustaches, a jagged white patch on the lower back, silky white flank tufts, and prominent white tips on the outer pairs of tail feathers. The sexes are similar, and young birds resemble adults except for an entirely dusky gorget lacking red iridescence. The three subspecies vary in range and in the color intensity of the gorget and underparts. Long wings create a swift-like appearance as the birds swoop and glide in pursuit of flying insects, and a long bill enables them to extract nectar from large flowers such as coralbeans (*Erythrina*) and ceibas (*Ceiba*). Northern populations are migratory, leaving their breeding areas during the cooler months, and there is some post-breeding movement in response to flowering cycles even in the tropics.

DISTRIBUTION Subsp. *constantii* ranges from El Salvador to southern Costa Rica; subsp. *leocadiae* occurs in western and southwestern Mexico and western Guatemala; subsp. *pinicola* is a resident of northwestern Mexico that has been spotted it the southwestern United States

HABITAT Lowland tropical deciduous, evergreen, and riparian forest, thorn scrub, savanna, plantations; lower mountain woodlands post-breeding; 0–5,600 ft (0–1,700 m)

SIZE Length: 4½–5⅜ in (11.5–13.5 cm). Weight: 7–9 g

STATUS Least Concern

Purple-collared Woodstar

This tiny hummingbird has a long, downcurved bill and is mainly shining olive green. The male has a turquoise throat patch bordered with a violet collar, a white belly, and a deeply forked tail. The female's underparts are entirely light orange-buff, and her round-ended tail has white tips to the outer feathers. The two subspecies are subsp. *fanny* and the longer-tailed subsp. *megalura*. A traplining species, the bird forages along a predictable route, visiting flowers of cacti, firecracker plants (*Russelia*), and other low-growing shrubs. The female builds a tiny cup nest on a thin, forking branch, usually at least 20 feet (6 m) above the ground, and incubates her two eggs for 15 to 16 days. The young are female-like in plumage and tail shape, and fledge at 19 to 22 days. Some post-breeding altitudinal movement is likely, though has not yet been confirmed. This is a common bird throughout its range.

DISTRIBUTION Subsp. *fanny* occurs in west and southeast Ecuador, and west Peru; subsp. *megalura* occurs in northern Peru

HABITAT Dry coastal scrub, woodland, gardens; 0–9,850 ft (0–3,000 m)

SIZE Length: 3–3⅛ in (7.5–8 cm). Weight: 2–2.5 g

STATUS Least Concern

Rufous-shafted Woodstar

The male Rufous-shafted Woodstar has a fairly long, forked tail and dark green plumage with a neat violet throat patch and broad white breast band. The female is bronzy green above and rich cinnamon buff below. There are three subspecies: subsp. *jourdanii*, described above; subsp. *rosae*, which has a lighter, almost orange-toned throat patch; and subsp. *andinus*, with a redder throat patch. The species watches for feeding opportunities from a high perch, but typically visits flowers at low to mid-levels in the strata. As well as discreetly entering the territories of other species, it will also visit small flowers that are overlooked by most other hummingbird species; it also takes many tiny insects, catching them in flight. It is usually rather uncommon (though as with other woodstars is easily overlooked) but is thought not to be in decline. It is, however, said to be very rare in Trinidad.

DISTRIBUTION Subsp. *jourdanii* occurs in Trinidad and northeast Venezuela; subsp. *rosae* occurs in the highlands of north Venezuela; subsp. *andinus* occurs in the Andes of Venezuela and the eastern Andes in northeast Colombia

HABITAT Forest edges, scrubland, plantations; 2,950–9,850 ft (900–3,000 m)

SIZE Length: 2⅜–3⅛ in (6–8 cm). Weight: unknown

STATUS Least Concern

Slender-tailed Woodstar

The male of this hummingbird has a long, narrow tail, and elongated violet feathers that flare out from the throat sides. The plumage is otherwise bright green above and white below with green speckles, and the face and throat are blackish. The female has a short, cinnamon-sided tail, green upperparts, and whitish-buff underparts. The male's eclipse plumage is a pale throat. Although this is the only species in its genus, the Slender-tailed Woodstar is very closely related to the *Calliphlox* woodstars. It is an inconspicuous species with a buzzing, insect-like flight and notably quick wing beats. It usually feeds alone at higher levels in the forest strata. It is not territorial and will be chased from other species' territories if they detect it. Very little is known of its ecology, and although it is not thought to be in decline, a thorough survey of its range would be helpful to establish its status.

DISTRIBUTION Central and south Bolivia to northern Argentina

HABITAT Thickly vegetated ravines and slopes, patchy deciduous forest; 5,250–8,550 ft (1,600–2,600 m)

SIZE Length: 2¾–3½ in (7–9 cm). Weight: unknown

STATUS Least Concern

Slender Sheartail

This elegant hummingbird has a fairly long, downcurved bill, and the male has a very long, narrow, deeply forked black tail. The male's plumage is mainly glossy, rich green, with a broad iridescent throat patch, bluish on the chin and shading to violet towards the breast. It also has a white breast band and undertail. The plainer female is olive above and warm cinnamon brown below, and has a short green tail with orange sides and white tips. The species feeds mainly on low-growing flowers, sometimes almost at ground level, and females are noted for restlessly wagging and fanning their tails while feeding, a behavior not noted in the males, which hold their tail closed and almost vertical while feeding. Both sexes frequently give a fast, dry call, both in flight and while perched. This bird is rather little-known but is thought to be declining as its habitat is rapidly being converted for agricultural use.

DISTRIBUTION South Mexico to Guatemala, Honduras and El Salvador

HABITAT Forest edges and clearings, secondary growth, lighter woodland and scrub; 3,300–9,850 ft (1,000–3,000 m)

SIZE Length: 3⅛–4⅞ in (8–12.5 cm). Weight: 2.5 g

STATUS Least Concern

Mexican Sheartail

Sparkling-tailed Hummingbird

This species is much shorter-tailed than the related Slender Sheartail. The male has a dark forked tail with an orange center, an entirely white breast and belly, a violet throat patch, and brown-tinged green upperside plumage. The female has green upperparts and an off-white underside. This hummingbird has similar feeding behavior to the Slender Sheartail. The nest is built low in a tree or shrub and is a small cup made mainly of plant seeds, anchored to a branch fork. The two populations show distinctly different courtship behavior—Yucatán males show a swooping display flight, while Veracruz males hover before the female, flaring the throat patch. The Yucatán population (about 6,000 to 10,000 birds) is restricted to the coastline, while the Veracruz population (no more than 2,500 birds) reaches farther inland. Both populations are seriously threatened by habitat loss.

The male of this very small, short-billed hummingbird (sometimes known as the Sparkling-tailed Woodstar) is strikingly marked, with a long, forked tail banded with black and white, a broad white breast band and flank spot, and a glossy violet-blue throat. The plumage is otherwise dark, shining green, shading to brownish on the back. The short-tailed female is green above (brightest on the back and tail) and has russet-buff underparts. The species is a trapliner with a rather slow, weaving flight, and visits a very wide range of flower types. It is also an avid flycatcher, making long sallies from high perches to chase its prey. This distinctive hummingbird is rather little-known and its nest and much of its breeding behavior remain undescribed by science, although displaying males are said to spread and close the tail repeatedly. The population is probably stable.

DISTRIBUTION Southeast Mexico, in two well-separated populations (Veracruz and Yucatán Peninsula)

HABITAT Dry scrub, mangrove edges, gardens, dry forest; 0–4,600 ft (0–1,400 m)

SIZE Length: 3⅜–3⅞ in (8.5–10 cm). Weight: 2.5 g

STATUS Near Threatened

DISTRIBUTION Central and southern Mexico, south through Guatemala, El Salvador, and Honduras into northern Nicaragua

HABITAT Dry forest and forest edges, oak woodland, thick scrub, secondary growth; 1,650–8,200 ft (500–2,500 m)

SIZE Length: 2½–3⅞ in (6.5–10 cm). Weight: 2.5–3 g

STATUS Least Concern

Beautiful Hummingbird

Glow-throated Hummingbird

The Beautiful Hummingbird is poorly known, in part because of confusion with its more widespread and nearly identical northern relative, the Lucifer Hummingbird. The genus name comes from Greek roots meaning "beautiful chest," referring to the long rose-purple gorget of the adult male. *Calothorax* are closely related to the sheartails (genus *Doricha*), and adult males sport shorter versions of the deeply forked scissor-like tails that give that group its name. As in the Lucifer Hummingbird, females and young males have creamy underparts washed with cinnamon and shorter, less forked tails banded in rufous, green, black, and white. During the nesting season, the male Beautiful Hummingbird performs a dramatic courtship display for the perched female, shuttling back and forth with bills touching before climbing high into the air for a high-speed dive directly over the female.

This very small, compact hummingbird has a rather short bill. The male is bright, shining green with a glittering pinkish-red throat patch and a wide white breast band below, while the female's upperparts are more olive yellow, and her throat is white with heavy, dark spotting. This species has a very restricted distribution in areas that are not much studied by ornithologists, hence its habits are very little known and poorly understood. Its behavior is, however, probably broadly similar to that of the other *Selasphorus* species, especially the closely related Scintillant Hummingbird. The Glow-throated is known to occur at six separate sites, some of which are at least partly protected, but it is by no means easy to find and appears to be continuing to decline. The estimated population is no more than 1,500 individuals and may number as few as 250. Further surveys on its distribution are required.

DISTRIBUTION Southern Mexico

HABITAT Thorn scrub, tropical deciduous forest; 3,300–7,200 ft (1,000–2,200 m)

SIZE Length: 3⅛–3½ in (8–9 cm). Weight: 3 g

STATUS Least Concern

DISTRIBUTION Western and central Panama

HABITAT Forest glades and edges; 2,500–5,900 ft (750–1,800 m)

SIZE Length: 2¾ in (7 cm). Weight: unknown

STATUS Vulnerable

Further Reading

Key general titles

Numerous books about hummingbirds are available and more appear every year. The following general titles are particularly useful.

Burton, R. 2001. *The World of the Hummingbird*. Willowdale, Ontario: Firefly Books. (A good up-to-date, popular account of the biology of hummingbirds. Includes photographs of many species.)

Del Hoyo, J., A. Elliot, & J. Sargantal (eds). 1999. *Handbook of Birds of the World*, Vol. 5: Barn-owls to Hummingbirds. Barcelona: Lynx Edicions. (The most comprehensive treatment of the hummingbirds of the world.)

de Schauensee, Rodolphe Meyer, & William H. Phelps. 1978. *A Guide to the Birds of Venezuela*. Princeton, New Jersey: Princeton University Press.

Dickinson, E.C. & J.V. Remsen Jr (eds). 2013. *The Howard & Moore Complete Checklist of the Birds of the World*. 4th Edition, Vol. 1. Eastbourne, UK: Aves Press. (Includes the most up-to-date checklist of all hummingbirds.)

Greenewalt, C.H. 1960. *Hummingbirds*. Garden City, New York: Doubleday & Company. Reprinted 1990. New York: Dover Publications. (A general account including detailed descriptions of feather structure, iridescence, and flight. Includes photographs of many species.)

Mazariegos H., L.A. 2000. *Hummingbirds of Colombia*. Cali, Colombia: Published by the author. (Deals with Colombian habitats, hummingbird biology, and conservation issues. Includes many beautiful photographs.)

Skutch, A.F. 1973. *The Life of the Hummingbird*. New York: Vineyard Books. (A general account of the life of hummingbirds, written in the author's inimitable style.)

Williamson, S.L. 2002. *A Field Guide to Hummingbirds of North America* (Peterson Field Guide Series). Boston: Houghton Mifflin Co.

Other useful articles and books

Chavez-Ramirez, F. & M. Dowd. 1992. "Arthropod feeding by two Dominican hummingbird species." *Wilson Bulletin* 104:743–747.

Clark, C.J., & T.J. Feo. 2008. "The Anna's hummingbird chirps with its tail: a new mechanism of sonation in birds." *Proceeding of the Royal Society of London* B 275:955–962.

Dalsgaard, Bo, Allan Timmermann, Ana M. Martin Gonzalez, Jens M. Olesen, Jeff Ollerton, & Laila H. Andersen. 2010. "*Heliconia*–hummingbird interactions in the Lesser Antilles: A geographic mosaic?" *Caribbean Journal of Science* 46(2–3): 328–331.

Davis, T.A.W. 1958. "The displays and nests of three forest hummingbirds of British Guiana." *Ibis* 100:31–39.

Diaz, L., & A.A. Cocucci. 2003. "Functional gynodioecy in *Opuntia quimilo* (Cactaceae), a tree cactus pollinated by bees and hummingbirds." *Plant Biology* 5(5):531–539.

Fjeldså, Jon, & Niels Krabbe. 1990. *Birds of the High Andes: A Manual to the Birds of the Temperate Zone of the Andes and Patagonia, South America*. Copenhagen: Zoological Museum, University of Copenhagen and Apollo Books.

Gill, Frank B. 1985. "Hummingbird flight speeds." *The Auk* 102(1):97–101.

Gould, W., C. Alarcón, B. Fevold, M.E. Jiménez, S. Martinuzzi, G. Potts, M. Solórzano, & E. Ventosa. 2007. *Puerto Rico Gap Analysis Project – Final Report*. Río Piedras, Puerto Rico: USGS, Moscow ID and the USDA FS International Institute of Tropical Forestry. 159 pp. and 8 appendices.

Graves, G.R. 1999. "Taxonomic notes on hummingbirds (Aves: Trochilidae), 2. *Popelairia letitiae* (Bourcier & Mulsant, 1852) is a valid species." *Proceedings of the Biological Society of Washington* 112:804–812.

Graves, G.R. & S.L. Olson. 1987. "*Chlorostilbon bracei* Lawrence, an extinct species of hummingbird from New Providence Island, Bahamas." *The Auk* 104:296–302.

Hammond, E.A. 1963. "Dr. Strobel's account of John J. Audubon." *The Auk* 80(4):462–466.

Hartman, Frank A. 1954. "Cardiac and pectoral muscles of trochilids." *The Auk* 71(4):467–469.

Hilty, S.L., & W.L. Brown. 1986. *A Guide to the Birds of Colombia*. Princeton, New Jersey: Princeton University Press.

Hu, D.-S., L. Joseph, & D. Agro. 2000. "Distribution, variation, and taxonomy of *Topaza* hummingbirds (Aves: Trochilidae)." *Ornitologia Neotropical* 11:123–142.

Hume, Julian P., & Michael Walters. 2012. *Extinct Birds* (Poyser Monographs). New York: Bloomsbury Publishing.

Kirchman, J.J., C.C. Witt, J.A. McGuire, & G.R. Graves. 2010. "DNA from a 100-year-old holotype confirms the validity of a potentially extinct hummingbird species." *Biology Letters* 6(1):112–115.

Kodric-Brown, Astrid, James H. Brown, Gregory S. Byers, & David F. Gori. 1984. "Organization of a tropical island community of hummingbirds and flowers." *Ecology* 65(5): 1358–1368.

Land, Hugh C. 1970. *Birds of Guatemala*. Wynnewood, Pennsylvania: Livingston Publishing Co.

Larkin, Claire C., Charles Kwit, Joseph M. Wunderle Jr., Eileen H. Helmer, M. Henry H. Stevens, Montara T.K. Roberts, & David N. Ewert. 2012. "Disturbance type and plant successional communities in Bahamian dry forests." *Biotropica* 44(1):10–18.

McGuire, Jimmy A., Christopher C. Witt, J.V. Remsen Jr, R. Dudley, Douglas L. Altshuler. 2008. "A higher-level taxonomy for hummingbirds." *Journal of Ornithology* 150:155–165.

Miller, Richard S. 1985. "Why Hummingbirds Hover." *The Auk* 102(4):722–726.

Nicholson, E.M. 1931. "Field-notes on the Guiana king hummingbird." *Ibis* 13:534–553.

Oniki, Yoshika. 1996. "Band sizes of southeastern Brazilian hummingbirds." *Journal of Field Ornithology* 67(3):387–391.

Raffaele, Herbert, James Wiley, Orlando Garrido, & Janis Raffaele. 1998. *A Guide to the Birds of the West Indies*. Princeton, New Jersey: Princeton University Press.

Ridgely, R.S. 1976. *A Guide to the Birds of Panama*. Princeton, New Jersey: Princeton University Press.

Ridgely, R.S., & P.J. Greenfield. 2001. *The Birds of Ecuador*. Vol. II. Ithaca, New York: Cornell University Press.

Ridgway, Robert, & Herbert Friedmann. 1911. *The Birds of North and Middle America*, Volume 50, Issue 5. US Government Printing Office.

Snow, Barbara K. 1973. "The behavior and ecology of hermit hummingbirds in the Kanaku Mountains, Guyana." *Wilson Bulletin* 85(2):163–177.

Stiles, F. Gary. 1995. "Behavioral, ecological and morphological correlates of foraging for arthropods by the hummingbirds of a tropical wet forest." *Condor* 97(4):853–878.

Stiles, F. Gary, & Alexander F. Skutch. 1989. *A Guide to the Birds of Costa Rica*. Ithaca, New York: Cornell University Press.

Temeles, E.J., Y.J. Rah, J. Andicoechea, K.L. Byanova, G.S.J. Giller, S.B. Stolk, & W.J. Kress. 2013. "Pollinator-mediated selection in a specialized hummingbird–Heliconia system in the Eastern Caribbean." *Journal of Evolutionary Biology* 26: 347–356.

Vigle, Gregory O. 1982. "A nest of *Eutoxeres aquila heterura* in western Ecuador." *The Auk* 99(1):172–173.

Weller, A.-A. 1999. "On types of trochilids in The Natural History Museum, Tring II. Re-evaluation of *Erythronota* (?) *elegans* Gould 1860: a presumed extinct species of the genus *Chlorostilbon*." *Bulletin of the British Ornithologists' Club* 119: 197–202.

Wolf, Larry L. 1975. "Female territoriality in the Purple-Throated Carib." *The Auk* 92:511–522.

Zimmerman, Dale A. 1973. "Range expansion of Anna's Hummingbird." *American Birds* 27(5):827–835.

Useful websites

http://www.avibase.bsc-eoc.org
A world bird database hosted by Bird Studies Canada, the Canadian co-partner of BirdLife International.

http://www.birdlife.org
The official website of BirdLife International, the world's largest nature conservation partnership. It is a growing group of 121 partner organizations from around the world, one per country. BirdLife is the IUCN Red List authority for birds, classifying species in terms of the risk of extinction. This is a great resource, and gives the status and distribution of every hummingbird species.

http://ibc.lynxeds.com
The Internet Bird Collection. A growing resource of videos, photographs, and sounds of the world's birds, with detailed taxonomic information for every species.

http://neotropical.birds.cornell.edu
Cornell Lab of Ornithology: Neotropical Birds Online. A growing resource giving facts, figures, and photographs of the birds of the region, including hummingbirds.

http://www.wikiaves.com.br
A website dedicated to the birds of Brazil, which contains very many hummingbird descriptions, maps, photographs, and sounds.

Index of Common Names

Index of Common Names

Index of Scientific Names

Index of Scientific Names

Michael Fogden and his wife, Patricia, have spent the past 35 years as freelance wildlife writers and photographers. Between them they have written 11 books and numerous articles, and in 2008 they received a Lifetime Achievement Award from the North American Nature Photography Association (NANPA).

Marianne Taylor is a freelance writer, editor, and illustrator. She has worked as editor for the bird book publisher Christopher Helm and a subeditor on the magazine *Birdwatch* and has written more than a dozen books on wildlife and science. She is also a keen wildlife photographer.

Sheri L. Williamson is a naturalist, ornithologist, award-winning writer, and author of *A Field Guide to Hummingbirds of North America*. Since 1996, she has worked as director of the Southeastern Arizona Bird Observatory. In addition to her field work, Sheri leads birding and nature tours and speaks at birding events.

MICHAEL FOGDEN I am greatly indebted to the many researchers who have studied and written about the biology and the natural history of hummingbirds. I must also thank everyone connected with Ivy Press with whom I have worked on this project: Jason Hook, Tom Kitch, James Lawrence, David Price-Goodfellow, Susi Bailey, and the two authors of the species accounts, Marianne Taylor and Sheri Williamson.

MARIANNE TAYLOR I would like to thank Jason Hook at Ivy Press for asking me to become involved with this book; Tom Kitch, who oversaw the process of creating the book from start to finish; and my coauthor, hummingbird expert Sheri Williamson, for working with me on such an interesting project. Our consultant (and author of the introductory material), Michael Fogden, provided invaluable feedback throughout, and I would also like to thank Susi Bailey for her superb copy-editing and helpful criticism. Many photographers have contributed their work to this book, creating between them the most dazzling collection of hummingbird images ever published. Without their skill, patience, and vision, this book would not have been possible. It was a great pleasure to work with David Price-Goodfellow as project manager, and thanks to James Lawrence and the Lanaways, who put together the wonderful layout that shows off these spectacular birds to their best possible advantage. My thanks also go to our proofreader Claire Saunders. Finally, I would like to thank the many dedicated ornithologists and fieldworkers who are committed to understanding hummingbird species (and, in many cases, saving them from extinction). Their work is crucial to the future of some of the world's most beautiful and remarkable birds.

SHERI L. WILLIAMSON My sincere appreciation goes to my husband and colleague, Tom Wood, for his patience and support, my coauthors and the team at Ivy Press, who did a remarkable job under extremely challenging conditions, the international community of hummingbird researchers and enthusiasts, past and present, whose passion and dedication have broadened and deepened our understanding of these remarkable creatures, and, of course, the birds that inspire us.

Ivy Press would like to thank the following for permission to reproduce copyright material:
Anselmo d'Affonseca (44, 286), Roger Ahlman (115, 321), Nick Athanas (69, 70, 89, 90, 113, 128, 129, 133, 153, 160, 186, 210, 216, 219, 237, 236, 251, 272, 275, 287, 300, 303, 318, 319, 327), Claude Balcaen/Biosphoto/FLPA (51, 81, 99), Walter Baliero (121), Glenn Bartley (back cover lower-center and bottom, 48, 60, 101, 103, 114, 122, 123, 130, 135, 138, 139, 141, 142, 143, 156, 158, 190, 201, 277), Rolf Bender/Imagebroker/FLPA (148), Neil Bowman/FLPA (80, 198, 208, 215, 278, 288), Eleanor Briccetti (45, 59, 281), Dusan Brinkhuizen (125, 147, 167, 249), Jose Cañas (312), Luis Casiano/Biosphoto/FLPA (82), Jean-Paul Chatagn/Biosphoto/FLPA (258), Robin Chittenden/FLPA (73, 98, 213, 259), Murray Cooper (253), Murray Cooper/Minden Pictures/FLPA (33, 38, 83, 88, 116, 117, 140, 146, 152, 154, 163, 165, 247, 256, 320), Sylvain Cordier/Biosphoto/FLPA (41, 222), Luiz Damasceno (39, 46, 68, 199, 301), Ian Davies (168, 169, 221, 338, 339), O. Deutsch/VIREO (239), John van Dort (265, 304), Knut Eisermann (223, 305, 311, 340), Kevin Elsby/FLPA (333), Marc Fasol (52), Tim Fitzharris/Minden Pictures/FLPA (207, 345), Michael and Patricia Fogden (front cover, spine, back cover top-right and upper-centre, 2, 3, 8, 10, 10, 11, 12, 13, 14, 15, 16, 17, 18, 19, 21, 23, 25, 25, 26, 26, 27, 27, 28, 30, 34, 35, 37, 40, 42, 47, 53, 54, 55, 58, 61, 62, 63, 65, 66, 67, 71, 74, 75, 85, 86, 87, 94, 95, 96, 97, 104, 105, 106, 107, 109, 110, 111, 126, 127, 136, 137, 145, 151, 159, 161, 162, 166, 170, 171, 173, 177, 178, 179, 180, 181, 182, 183, 184, 185, 189, 191, 193, 194, 195, 196, 197, 203, 204, 205, 214, 218, 224, 225, 226, 227, 229, 230, 231, 232, 233, 235, 238, 240, 241, 242, 244, 252, 255, 260, 262, 263, 264, 268, 274, 280, 284, 285, 295, 297, 307, 308, 309, 310, 315, 316, 317, 324, 325, 326, 348, 349, 350, 351), Markus Friedrich/Imagebroker/FLPA (36), Steve Gettle/Minden/FLPA (76), Michael Gore (323), Wim de Groot (57), Gregory Guida/Biosphoto/FLPA (131, 155, 248), Peter Hodum (92, 93), Lee Hunter (332), Imagebroker/FLPA (119, 314), Oláh János (134, 157, 246, 273), Donald M. Jones/Minden Pictures/FLPA (342), Mike Lane/FLPA (43, 102, 217, 228, 261), Osvaldo Larraín (120), Damien Laversanne/Biosphoto/FLPA (257, 271), Chris Llewellyn (290), Peter Llewellyn/FLPA (299), James Lowen/FLPA (77, 209, 245, 313), Claus Meyer/Minden Pictures/FLPA (50), Pete Morris (32, 49, 64, 150, 187, 269, 282, 283, 289, 293), Nature Photographers Ltd./Brian Small (343), Alex Navarro (267), Rolf Nussbaumer/Imagebroker/FLPA (78, 79, 279, 331, 330, 335, 341), Chechi Peinado (302), Photo Researchers/FLPA (72, 174, 254, 270, 291, 334, 337, 347), Tui De Roy/Minden Pictures/FLPA (175), Dubi Shapiro (164, 250), Glen Tepke (266), Roger Tidman/FLPA (243), David Tipling/FLPA (1, 100, 206), Daniel Uribe (132, 144, 149, 211), Raul Vega (234), Tom Vezo/Minden Pictures/FLPA (212, 306), Jay Warburton (56), Egon Wolf Miranda (91), and Konrad Wothe/Minden Pictures/FLPA (329).

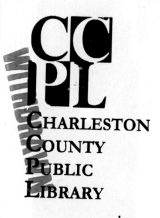